Sustainable Manufacturing for Industry 4.0

Manufacturing Design and Technology Series

Series Editor:
J. Paulo Davim,
University of Aveiro, Portugal

This series will publish high quality references and advanced textbooks in the broad area of manufacturing design and technology, with a special focus on sustainability in manufacturing. Books in the series should find a balance between academic research and industrial application. This series targets academics and practicing engineers working on topics in materials science, mechanical engineering, industrial engineering, systems engineering, and environmental engineering as related to manufacturing systems, as well as professions in manufacturing design.

Drills: Science and Technology of Advanced Operations
Viktor P. Astakhov

Technological Challenges and Management: Matching Human and Business Needs
Edited by Carolina Machado and J. Paulo Davim

Advanced Machining Processes: Innovative Modeling Techniques
Edited by Angelos P. Markopoulos and J. Paulo Davim

Management and Technological Challenges in the Digital Age
Edited by Pedro Novo Melo and Carolina Machado

Machining of Light Alloys
Aluminum, Titanium, and Magnesium
Edited by Diego Carou and J. Paulo Davim

Additive Manufacturing: Applications and Innovations
Edited by Rupinder Singh and J. Paulo Davim

Emotional Intelligence and Neuro-Linguistic Programming: New Insights for Managers and Engineers
Edited by Carolina Machado and J. Paulo Davim

Business Intelligence and Analytics in Small and Medium Enterprises
Edited by Pedro Novo Melo and Carolina Machado

Enabling Technologies for the Successful Deployment of Industry 4.0
Edited by Antonio Sartal, Diego Carou and J. Paulo Davim

Industry 4.0: Challenges, Trends, and Solutions in Management and Engineering
Edited by Carolina Machado and J. Paulo Davim

Sustainable Manufacturing for Industry 4.0: An Augmented Approach
Edited by K. Jayakrishna, Vimal K.E.K., S. Aravind Raj, Asela K. Kulatunga, M.T.H. Sultan and J. Paulo Davim

Functional and Smart Materials
Edited by Chander Prakash, Sunpreet Singh and J. Paulo Davim

Advanced Manufacturing and Processing Technology
Edited by Chander Prakash, Sunpreet Singh, and J. Paulo Davim

Characterization, Testing, Measurement, and Metrology
Edited by Chander Prakash, Sunpreet Singh, and J. Paulo Davim

For more information about this series, please visit: https://www.crcpress.com/Manufacturing-Design-and-Technology/book-series/CRCMANDESTEC

Sustainable Manufacturing for Industry 4.0

An Augmented Approach

Edited by
K. Jayakrishna, Vimal K.E.K., S. Aravind Raj,
Asela K. Kulatunga, M.T.H. Sultan, and
J. Paulo Davim

CRC Press
Taylor & Francis Group
Boca Raton London New York

CRC Press is an imprint of the
Taylor & Francis Group, an **informa** business

First edition published 2020
by CRC Press
6000 Broken Sound Parkway NW, Suite 300, Boca Raton, FL 33487-2742

and by CRC Press
2 Park Square, Milton Park, Abingdon, Oxon, OX14 4RN

© 2021 Taylor & Francis Group, LLC
CRC Press is an imprint of Taylor & Francis Group, an Informa business

Library of Congress Cataloging-in-Publication Data
Names: Jayakrishna, K., 1984- editor.
Title: Sustainable manufacturing for industry 4.0 : an augmented approach /
edited by K. Jayakrishna, Vimal KEK, S. Aravind Raj, K.M.A.K. Kulatunga,
Mohamed Thariq Bin Haji Hameed Sultan and J. Paulo Davim.
Description: First edition. I Boca Raton : CRC Press, 2020. I Series:
Manufacturing design & technology I Includes bibliographical references
and index.
Identifiers: LCCN 2020015133 (print) I LCCN 2020015134 (ebook) I ISBN
9781138606845 (hbk) I ISBN 9780429466298 (ebk)
Subjects: LCSH: Manufacturing processes--Environmental aspects. I
Engineering design--Environmental aspects. I Manufacturing
industries--Technological innovations. I Manufacturing
processes--Automation.
Classification: LCC TS155.7 .S858 2020 (print) I LCC TS155.7 (ebook) I
DDC 670--dc23
LC record available at https://lccn.loc.gov/2020015133
LC ebook record available at https://lccn.loc.gov/2020015134

ISBN: 978-1-138-60684-5 (hbk)
ISBN: 978-0-429-46629-8 (ebk)

Typeset in Times
by Deanta Global Publishing Services, Chennai, India

Contents

Foreword

Globalisation has created a huge demand for capital and consumer goods that organisations worldwide need to meet, but without compromising the sustainable development of societies and without endangering the natural ecosystem. To address this significant challenge, organisations, and especially manufacturing companies, must create value which is aligned to the social, environmental and economic pillars of sustainability.

On the other hand, most manufacturing industries and companies have now embarked on the transition towards the adoption of digital technologies, which has given rise to the fourth stage of industrialisation, the so-called Industry 4.0. Presently, Industry 4.0 and its associated technologies, like internet of things (I.o.T.), Big Data and data analytics, additive manufacturing or 3D printing, advanced robotics, augmented and virtual reality, cloud computing, simulation, machine learning and artificial intelligence, among others, are asserting a substantial influence on the manufacturing sector. Additionally, new business models are being developed around these technologies. The adoption of Industry 4.0 technologies and development of new business models suitable to facilitate the digitalisation of manufacturing systems and operations provide enormous opportunities for the realisation of sustainable manufacturing capable of ensuring the balanced social, environmental and economic evolution of human societies without compromising the existence of future generations.

This book contains valuable contributions from academics, researchers and industrial experts who have been involved with, or have worked in, the manufacturing sector and who fervently believe that Industry 4.0 can contribute to make the manufacturing industry a model of sustainability. In my own experience, company failures to effectively adopt Industry 4.0 technologies and make manufacturing operations more sustainable derive from the lack of core knowledge about these and how they interact with and influence each other. In this regard, this book presents an invaluable opportunity for academics, researchers and industrialists to obtain deep and detailed knowledge on sustainable manufacturing for Industry 4.0 considered from an augmented approach and perspective. To accomplish this, the book is divided into six main sections, one that provides an understanding of the general context of Industry 4.0 (Section 1); one that delves into the relationship of sustainable manufacturing and Industry 4.0 (Section 2); one that focuses on discussing the innovations that enable smart factories (Section 3); one that shows how Industry 4.0 technologies can contribute to effective sustainable decision-making (Section 4); one which relates the concept of smart factories and aligns the characteristics of these places to comply with environmental regulations and aspects (Section 5); and the final section (Section 6), which presents a framework, some case studies and other elements that need to be considered to ensure that Industry 4.0 companies are also sustainable. By focusing on these fundamental aspects, the readers of this book will:

- Gain knowledge of the fundamentals of Industry 4.0 and its potential contribution to making the manufacturing industry, and its systems and operations, more sustainable;
- Understand how to adapt Industry 4.0 technologies and their new and disruptive business models to contribute towards the sustainable development of manufacturing systems and operations;
- Reflect upon and consider particular operational issues that have a direct impact on the digitalisation of manufacturing companies when adopting Industry 4.0 technologies and gearing these towards the sustainability enhancement of their systems and operations.

I am very sure that this book will serve and provide an excellent read and reference to academics, researchers, all kind of business leaders and all those interested in sustainable manufacturing operations and Industry 4.0.

Jose Arturo Garza-Reyes
Head of the Centre for Supply Chain Improvement
College of Business, Law and Social Sciences
University of Derby
United Kingdom

Preface

Over the past four decades, sustainability has increasingly commanded the attention of business executives in many regions of the world. Growing numbers of companies are adopting socially responsible and environmentally conscious policies associated with a range of issues, such as climate change, water use, supply-chain management, responsible investment, equitable labour relations and preservation of resources and standards of living for future generations. Contemporary organisations are also applying creativity and innovation to the development of clean technologies that may yield startling increases in the productivity of the earth's resources and dramatic, broad-based improvements in living standards. Despite these developments, market shifts and disruptions are making the achievement of sustainable business models more difficult, while also making the issue more immediate. Surging industrial growth in emerging markets is placing a heavy strain on the supply of commodities and other non-renewable resources.

Industry 4.0 promises tremendous opportunities for industries to go green by leveraging the presence of virtual physical systems and internet-driven technologies to extend a competitive advantage and set the platform for the factory of the future and smart manufacturing. As manufacturing industries are entering the period of the fourth industrial revolution, which uses digital technologies exponentially to meet the increasing customer needs, the future factory is unfolding, with digital connectivity and advances in sustainable manufacturing. Sustainable product development lays the foundation for bringing about green/eco-products right from the design stage of product development. This book provides insights on Industry 4.0 to identify the potential opportunities underlying the sustainable manufacturing space by leveraging the internet of things (I.o.T.). This book contributes to the subject of sustainable product development and manufacturing by describing existing trends and methodologies adopted by modern digitalised industrial organisations that are satisfying economically and with societal benefits too, by presenting concepts for sustainable product designs, planning methods and manufacturing technologies. This book discusses a few strategies for knowledge dissemination on sustainability and feasible solutions employable within organisations aiming at implementing sustainability practices. In addition, this book explores the latest techniques that enable smartly connected manufacturing. In particular, how Industry 4.0 can assist continuous improvement in organisations with the help of the industrial internet of things (I.I.o.T.), and tackle past unsolved problems to increase sustainability and reliability.

This book offers phenomenal measures that can be adopted by practising design engineers to develop products that will be sustainable in all stages of their life-cycles. However, for successful implementation of sustainable manufacturing practices, formulation of strategies has been found to be critical. From this viewpoint, critical insights were presented by reviewing the existing practices of sustainable manufacturing by adopting the technological evolution from embedded systems to cyber-physical systems. This book provides insights on ways of deploying these practices in correlation with the environmental benefits planned to support the practising managers and stakeholders. This publication delivers research results which address the aforementioned challenges, solutions and implementation perspectives with regard to sustainable product development and manufacturing from an Industry 4.0 perspective.

Overall, this book provides insights into major fields of actions which are essential to make the earth worth living on, for now and forever, and for the future generations. Thus, finding solutions towards world-class manufacturing, which would concurrently consider the triple bottom line of sustainability is of utmost importance and more crucial than ever. This book also discusses how manufacturing companies can transform their business with the help of digital technology in the era of Industry 4.0 to drive intuitive assets and smarter manufacturing processes and optimise resources to enable sustainability in manufacturing.

Editors

K. Jayakrishna is an Associate Professor in the School of Mechanical Engineering at the Vellore Institute of Technology university in India. Dr Jayakrishna's research is focused on the design and management of manufacturing systems and supply chains to enhance efficiency, productivity, and sustainability performance. More recent research is in the area of developing tools and techniques to enable value creation through sustainable manufacturing, including methods to facilitate more sustainable product design for closed-loop material flow in industrial symbiotic set-up, and developing sustainable products using hybrid biocomposites. He has mentored undergraduate and graduate students (two M.Tech. Thesis and 24 B.Tech. Thesis) which have so far led to 40 journal publications in leading SCI/SCOPUS Indexed journals, 17 book chapters, and 85 refereed conference proceedings. Dr Jayakrishna's team has received numerous awards in recognition of the quality of the work that has been produced. He teaches undergraduate and graduate courses in the manufacturing and industrial systems area and his initiatives to improve teaching effectiveness have been recognized through national awards. He has also been awarded the Institution of Engineers (India) Young Engineer Award in 2019 and the Distinguished Researcher Award in the field of Sustainable Systems Engineering in 2019 by the International Institute of Organized Research.

Vimal K.E.K. received his Ph.D. degree in Sustainable Manufacturing in 2016 and his M.Tech degree in Industrial Engineering in 2012 from the National Institute of Technology, Tiruchirappalli, India, and a bachelor's degree in Production Engineering in 2010 from the PSG College of Technology, India. He has more than nine years of teaching and research experience in Mechanical Engineering, with special emphasis on sustainability and environmentally conscious manufacturing. He also has interests in operations and advanced manufacturing concepts, namely, lean and agile manufacturing. He has worked as a guest editor of the *International Journal of Enterprise Network Management* and *Polymers and Polymer Composites*. Presently, he acts as a reviewer for more than ten prestigious Web of Science journals. Besides, he has also published as the author (and co-author) more than eight book chapters and 40 articles in journals and conferences. He is currently working as Assistant Professor in the Department of Mechanical Engineering, National Institute of Technology, Patna, India.

S. Aravind Raj is currently working as Associate Professor in the School of Mechanical Engineering at the Vellore Institute of Technology, Vellore, India. He received his doctorate degree in Industrial Engineering from the National Institute of Technology, Tiruchirappalli in 2014. His research area is focused on agile manufacturing processes and their applications in manufacturing industries. Recent work focuses on Industrial Engineering, Additive Manufacturing, and Industry 4.0 projects. He has published 20 papers in leading SCI/SCOPUS Indexed journals and 34 papers in International conferences; he is co-author for four book chapters and has edited one book. He is a reviewer of reputable international journals. He has received several awards including a Doctoral Scholarship and Best Paper Award.

Asela K. Kulatunga is an academic, researcher, and consultant in the area of Sustainable Manufacturing, Climate Change mitigation and Green Logistics and Supply Chain Management, with over 13 years of experience in local and foreign assignments providing research, consultancy, and training services to the manufacturing sector and other local and foreign institutes. He is an expert in the areas of Life-Cycle Assessment, Eco-Innovation/Eco-Design, and Green Logistics and Supply Chain Management. He has more than eight years of experience in U.N.I.D.O. and U.N.E.P. initiated projects. He is a promoter of Sustainable Consumption and Production and an Accredited Professional of the Green Building Council of Sri Lanka and a Member of the European Roundtable for Sustainable Consumption and Production (M.E.R.S.C.P.).

M.T.H. Sultan is a Professional Engineer (P.Eng.) registered under the Board of Engineers Malaysia (B.E.M.) and a Chartered Engineer registered with the board of the Institution of Mechanical Engineers, United Kingdom, currently attached to Universiti Putra Malaysia as a Director of the Aerospace Manufacturing Research Centre (A.M.R.C.), Faculty of Engineering, U.P.M. Serdang, Selangor, Malaysia. Being a Director of A.M.R.C., he was also appointed as an Independent Scientific Advisor to the Aerospace Malaysia Innovation Centre (A.M.I.C.) based in Cyberjaya, Selangor, Malaysia. He received his Ph.D. from the University of Sheffield, United Kingdom. His area of research includes Hybrid Composites, Advanced Materials, Structural Health Monitoring, and Impact Studies. He has published more than 100 international journal papers. Currently, he an Associate Professor in the Department of Aerospace Engineering, Universiti Putra Malaysia.

J. Paulo Davim received a Ph.D. degree in Mechanical Engineering in 1997, an M.Sc. degree in Mechanical Engineering in 1991, a Mechanical Engineer degree (M.Eng.–5 years) in 1986, from the University of Porto (F.E.U.P.), the Aggregate title (Full Habilitation) from the University of Coimbra in 2005, and a D.Sc. from London Metropolitan University in 2013. He is Eur.Ing. in F.E.A.N.I.-Brussels and Senior Chartered Engineer with the Portuguese Institution of Engineers, with an M.B.A. and Specialist title in Engineering and Industrial Management. Currently, he is a Professor at the Department of Mechanical Engineering of the University of Aveiro, Portugal. He is the editor-in-chief of several international journals, guest editor of journals, books editor, book series editor, and scientific advisor for many international journals and conferences. Presently, he is an Editorial Board member of 25 international journals and acts as reviewer for more than 80 prestigious Web of Science journals. He has also published more than 110 books, 70 book chapters, and 400 articles in journals and conferences.

Contributors

Ramakurthi Veera Babu
Department of Mechanical Engineering
National Institute of Technology, Warangal
Warangal, India

Ashish Das
Department of Production and
    Industrial Engineering
National Institute of Technology
Jamshedpur, India

Someh Kumar Dewangan
Department of Computer Science and
    Engineering
RSR Rungta College of Engineering
    and Technology
Bhilai, India

Sumit Gupta
Department of Mechanical Engineering
Amity School of Engineering &
    Technology
Noida, India

K. Jayakrishna
School of Mechanical Engineering
Vellore Institute of Technology
Vellore, India

Srijit Krishnan
Department of Mechanical Engineering
Amity School of Engineering &
    Technology
Noida, India

Niraj Kumar
Department of Production and
    Industrial Engineering
National Institute of Technology
Jamshedpur, India

K. Lenin
Department of Mechanical Engineering
K. Ramakrishnan College of
    Engineering
Trichy, India

S. Uma Mageswari
Department of Business Administration
Auxilium College
Vellore, India

E. Manavalan
School of Mechanical Engineering
Vellore Institute of Technology
Vellore, India

Shambhu Kumar Manjhi
Department of Production and
    Industrial Engineering
National Institute of Technology
Jamshedpur, India

Vijaya Kumar Manupati
National Institute of Technology,
    Warangal
Warangal, India

J. Martinez-Girlado
Department of Industrial Engineering
EAM University Institution
Armenia, Colombia

Deepak Mathivathanan
Centre for Logistics and Supply Chain
    Management
Loyola Institute of Business
    Administration
Chennai, India

K. Mathiyazhagan
Department of Mechanical Engineering
Amity University, Noida
Noida, India

Sushil Kumar Maurya
Department of Mechanical Engineering
Modern Institute of Technology and
    Research Centre
Alwar, Rajasthan

M. Fakkir Mohamed
Department of Automobile Engineering
Tamilnadu College of Engineering
Coimbatore, India

Nikhil Wakode
Department of Mechanical Engineering
National Institute of Technology,
    Warangal
Warangal, India

M. Nishal
Department of Mechanical Engineering
Sri Venkateswara College of Engineering
Sriperumbudur, India

Shashi Bhusan Prasad
Department of Production and
    Industrial Engineering
National Institute of Technology
Jamshedpur, India

E. Vengata Raghavan
Accenture
Detroit, USA

Leos Safar
Faculty of Economics
Technical University of Kosice
Kosice, Slovakia

Prateek Saxena
Department of Mechanical Engineering
Technical University of Denmark
Lyngby, Denmark

Alokita Shukla
School of Mechanical Engineering
Vellore Institute of Technology
Vellore, India

Lokesh Singh
Department of Mechanical
    Engineering
RSR Rungta College of Engineering
    and Technology
Bhilai, India

Himanshu Singh
Department of Mechanical Engineering
National Institute of Technology
Nagpur, India

Sivakumar K.
Department of Mechanical
    Engineering
Siddharth Institute of Engineering and
    Technology (SIETK)
Puttur, Andhra Pradesh, India

A. Sofi
Department of Structural and
    Geotechnical Engineering
School of Civil Engineering
Vellore Institute of Technology
Vellore, India

Jakub Sopko
Faculty of Economics
Technical University of Kosice
Kosice, Slovakia

R. Subhaa
Faculty of Mechanical Engineering
SSM Institute of Engineering and
    Technology
Dindigul, India

R. Sudhakara Pandian
School of Mechanical Engineering
Vellore Institute of Technology
Vellore, India

Padmaja Tripathy
Department of Mechanical Engineering
National Institute of Technology
Rourkela, India

M.L.R. Varela
Department of Production Engineering
University of Minho
Braga, Portugal

Astha Vijaivargia
Department of Mechanical Engineering
University College of Engineering
Kota, India

Rahul Verma
School of Mechanical Engineering
Vellore Institute of Technology
Vellore, India

Vimal K.E.K.
Department of Mechanical Engineering
National Institute of Technology
Patna, India

S. Vinodh
Department of Production Engineering
National Institute of Technology
Tiruchirappalli, India

Vishal A. Wankhede
Department of Production Engineering
National Institute of Technology
Tiruchirappalli, India

Abdul Zubar Hameed
Department of Industrial Engineering
King Abdulaziz University
Jeddah, Saudi Arabia

1 Concept of Industry 4.0

1.1 INTRODUCTION AND EVOLUTION OF INDUSTRY 4.0

Deepak Mathivathanan

1.1.1 INTRODUCTION

The contemporary challenge we face today is to understand and cope with the newer and more sophisticated technological advances which have highly impacted the way we live (Lee et al., 2015). We are in an age of emerging technology breakthroughs where artificial intelligence, the Internet of Things, 3D printing, nanotechnology, blockchain technology, etc., are becoming ingrained in our day-to-day lives (Lee et al., 2014). Historically many such revolutions have occurred due to the introduction of newer technologies which triggered profound changes in our way of living in terms of economic and social structures (Rüßmann et al., 2015). Ever-changing markets, globalisation and the need for competitive advantage along with the technological advancements have led to the emergence of the conceptual idea of Industry 4.0 (Gorecky et al., 2014). Being one of the most disruptive revolutions, Industry 4.0 has become one of the most researched topics among professionals in industry and academia around the globe. The term 'Industry 4.0' was first introduced at the Hanover Messe trade fair, Germany in 2011 and was used to describe the ongoing digital transformation in global value chain creation (Hofmann & Rüsch, 2017). Germany Trade and Invest defines Industry 4.0 as 'A paradigm shift made possible by technological advances which constitute a reversal of conventional production process logic' (GTAI, 2014). In simple terms, it meant that machinery would no longer simply process the product, but the products could communicate with the machinery to instruct it exactly what to do. Klaus Schwab, in his article 'The Fourth Industrial Revolution', stated that 'Industry 4.0 includes business processes in industry that envisage organisation of global production networks on the basis of new information and communication technologies and Internet technologies, with the help of which interaction of the production objects is conducted' (Schwab, 2017). Loshkareva et al. (2015) defined Industry 4.0 as 'a revolutionary method of organisation of industrial production, based on wide digitization and automization of production and distribution processes in industry that erases limits between physical objects, turning them into a comprehensive complex system of interconnected and interdependent elements'. One must understand that the so-called Industry 4.0 is based on the concept of completely automated production capable of interacting autonomously with the main stakeholders.

Before going deep into explaining what Industry 4.0 is, it is also equally important to discuss the evolution of it with reference to history. As the term Industry 4.0

suggests, it is the fourth industrial revolution and there are three such revolutions registered earlier in history due to the radical developments in terms of transformation of how we manufacture. A brief overview of the four industrial revolutions in presented below.

   First Industrial Revolution: the invention of mechanisation and mechanical power generation in 1784 is marked as the first industrial revolution (Freeman & Soete, 1997). The introduction of power, and water and steam facilitating mechanisation of production led the transition from manual work to the first powered manufacturing processes (Brynjolfsson & McAfee, 2014). An improved quality of life which ensured the transition from manual labour to machine labour was the change offered and it continued for many decades (Landes, 2003). Though the invention of the steam engine marked the initial revolution in the late 17th century, the process of transformation from manual manufacturing to factory manufacturing using machines continued until the 19th century (Schwab, 2017). This revolution impacted the structure of society and led to urbanisation and the emergence of new technological developments. As a matter of fact, the invention of power made way for transportation around the globe and enhanced external trade which led nations in the West to transform themselves into global powerhouses (Klingenberg & do Vale Antunes, 2017).

   Second Industrial Revolution: the invention of electricity and the concept of production lines in the late 19th century kickstarted the second industrial revolution which extended until the early 20th century. Introduction of assembly line slaughterhouses in 1870 was one of the initial outcomes of the second revolution (Yin et al., 2018). The key role of the second industrial revolution was the replacement of steam by electricity, thanks to the invention of the first dynamo by E. Siemens in 1867. This was gradually followed by the invention of electric lamps, hydroelectric power stations, powerlines for transferring electricity and so on. Electricity-driven mass production, conveyor production of cars and other products are the major accomplishments of this revolution (Schwab, 2017). With developing industrialisation, there was a growing need for better communication technologies. The invention of the telegraph, underwater transcontinental telegraphic lines and wireless radio communication are part of this revolution. The eventual increase in labour efficiency and associated consequences that manifested in the later 19th century helped western countries to move towards a newer level of development (Klingenberg & do Vale Antunes, 2017). The socio-economic developments in western society accelerated at a greater pace than the rest of the world and hence those countries eventually turned into superpowers. The second industrial revolution reached the second wave countries as a case of forced modernisation only by the end of the 19th century (Xu et al., 2018).

   Third Industrial Revolution: manufacturers incorporating electronics and computer technology into their factories in the late 1950s marked the emergence of the third industrial revolution (Sommer, 2015). This was period when manufacturers began experiencing the disruptive shift in the way industries operate, with the introduction of more digital technologies and automation overpowering the traditional analogue and mechanical technologies (Yin et al., 2018). This is often referred to as a 'Digital Revolution' spurred by the electronics industry inventing the first programmable logic control system in 1969 (Kagermann et al., 2013). In the last few decades of the 20th century technological advances in terms of electronic devices, such as the transistor and, later, integrated circuit chips had begun to replace the operators with fully automated machines (Rifkin, 2016). These advances brought changes to manufacturing with the introduction of industrial robots and computer numeric controls (CNCs). This third industrial revolution also witnessed the development of software systems such as enterprise resources planning tools to track and plan inventory, and globalisation which enabled developed countries to go for low-cost manufacturing and assembly in low-cost countries (Deloitte, 2015). This extended formalising of logistics throughout the globe led to the concept of global supply chain management (Schwab, 2017). This third industrial revolution is still ongoing in a few parts of the world and so we cannot define the exact timescale of this revolution.

Fourth Industrial Revolution: the term 'Industry 4.0', also known as the fourth industrial revolution, emerged as a result of optimising the computerisation and addition of entirely new disruptive technologies introduced during the third revolution (Drath & Horch, 2014). To be precise, the initiation of Industry 4.0 was first officially publicised at the Hannover trade fair in 2011 under which it was intended to enhance the German competitive advantage from a manufacturing industry perspective (Schwab, 2017). The governmental bodies in Germany later included this idea in its 'High-Tech Strategy 2020' (Kagermann et al., 2013). Industry 4.0 is still in its infancy and aims at integrating manufacturing automation with customised and flexible mass production technologies. Industry 4.0 aims at providing a holistic manufacturing approach which emphasises digital technology with an advanced level of interconnectivity provided by the means of the Internet of Things (IoT) and cyber-physical systems making smart factories a reality (Lee et al., 2015). Cyber-physical systems make machines smarter with their ability to access real-time data, and help to share, analyse and guide intelligent actions of various processes in the industry (Hellinger & Seeger, 2011). Smart machines are capable of monitoring, detecting and predicting any errors dynamically which eventually results in increasing the production efficiency and productivity to greater levels. Cyber-physical systems even enable monitoring and managing industry operations from a remote location. Industry 4.0 can also be applied to aspects like quality control, capacity planning and production scheduling, and even logistics. Figure 1.1 depicts the four industrial revolutions along with the critical developments.

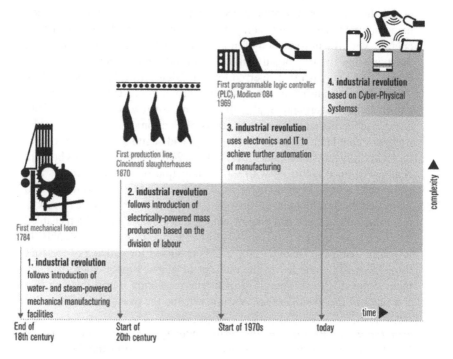

FIGURE 1.1   Evolution of Industry 4.0 (adopted from Kagermann et al., 2013).

1.1.2   WHAT IS INDUSTRY 4.0?

Industry 4.0 is nothing but the technological evolution of cyber-physical systems leaping forward from the era of embedded systems. According to a McKinsey report (2016), Industry 4.0 is

> the next phase in the digitisation of the manufacturing sector, driven by four disruptions: the astonishing rise in data volumes, computational power, and connectivity, especially new low-power wide-area networks; the emergence of analytics and business-intelligence capabilities; new forms of human-machine interaction such as touch interfaces and augmented-reality systems; and improvements in transferring digital instructions to the physical world, such as advanced robotics and 3-D printing.

The European Parliament defined Industry 4.0 as

> a group of rapid transformations in the design, manufacture, operation and service of manufacturing systems and products. The 4.0 designation signifies that this is the world's fourth industrial revolution, the successor to three earlier industrial revolutions that caused quantum leaps in productivity and changed the lives of people throughout the world
>
> **(Smit et al., 2016).**

The German Federal Ministry of Education and Research (2016) defined Industry 4.0 as:

> Industrie 4.0 combines production methods with state-of-the-art information and communication technology. The driving force behind this development is the rapidly increasing digitization of the economy and society. The technological foundation is provided by intelligent, digitally networked systems that will make largely self-managing production processes possible.

> Brettel et al. (2014) stated that 'Industry 4.0 focuses on the establishment of intelligent products and production processes'. Drath and Horch (2014) argued that Industry 4.0 is 'often understood as the application of the generic concept of cyber-physical systems (CPS)'.

Industry 4.0 integrates the production systems both horizontally (i.e., from supplier to customer) and vertically (shop floor to ERP level) based on real-time data exchange for customised flexible manufacturing (Deloitte, 2015). Such production systems can unite manufacturing facilities, smart machines and storage systems termed cyber-physical production systems, leading to the concept of smart manufacturing and smart factories. Though Industry 4.0 has received a lot of attention in recent years, the concept still hard to put into practice for many entrepreneurs due to the lack of a structured and systematic implementation model. But Industry 4.0 can boost productivity and reduce production costs without compromising on the quality of the products. Industry 4.0 can integrate business and manufacturing with the firm's suppliers and customers with the help of the Internet of Things and cyber-physical systems (Ardanza et al., 2019). The Internet of Things and cyber-physical systems serve as the building blocks that provide decentralised control and connectivity that sophisticatedly share real-time information to identify, locate, track, monitor and optimise any production processes (Bagheri et al. 2015). The integral components of Industry 4.0 are the cyber-physical systems. These systems serve as a bridge between the digital and physical components of the industry with the help of internet technology. These systems comprise embedded software and communication capabilities which provide intelligent control systems which can be connected in a network of such similar systems. Cyber-physical systems incorporate advanced mechatronics and adaptronics-enabled sensors, controllers and actuators which can be uniquely identified with an Internet Protocol (IP) address. These cyber-physical systems enable building of smart factories, smart logistics and other smart applications (Bag et al., 2018). For complete transformation towards Industry 4.0, organisations should incorporate the supportive technologies which are the components. The following sections detail the components and benefits of the Industry 4.0 transformation.

1.1.3 Components of Industry 4.0

There has been a lot of buzz among manufacturers and industrialists about Industry 4.0, as they see it as a solution to address the issues in organisational communications and transform the manufacturing and production process from traditional to smart factory level. To enable such a transformation and to reap the future benefits of Industry 4.0, it is necessary to distinguish its main components. The details of the nine main components are presented in the following subsections aimed at building a brief understanding of them.

(i) *Big Data and Analytics*: Big data analytics has evolved as a more reliable and powerful tool to solve value chain problems. With existing advanced technological capabilities, an industrial venture with software systems to handle huge volumes of data can develop innovative business models based on the data resources (Yin & Kaynak, 2015). Big data analytics involve analyses of large datasets of information about customers preferences, correlations, market trends and other information to benefit the operations, marketing and customer experience for a business (Seele, 2017). These concepts of big data can be applied in industries to accelerate their competitive advantage by solving the challenges at the organisational level, measuring and monitoring towards increasing productivity and innovation (Hermann et al., 2016).

(ii) *The Industrial Internet of Things*: The concept of IoT that is embedded in Industry 4.0 and is very similar to the concept of cyber-physical systems that is referred to as the Industrial Internet of Things (IIoT). IoT provides the platform for objects and devices such as sensors, actuators and even cell phones to communicate with other devices and machinery, as well as humans, to identify, report and work out solutions for any protocol (Kagermann et al., 2013). IIoT the cloud-based technological revolution aimed at providing computational and analytical solutions. The integration of such technology allows objects to work and solve problems in real time, independently, with minor intervention from human force (Rahman & Rahmani, 2018).

(iii) *Cloud Computing*: The other most important component of Industry 4.0 is the cloud. The cloud refers to the many IT resources which offer virtual storage and processing capabilities to multiple users. Rahman et al. (2018) emphasised that cloud computing provides the platform for technologies like Internet of Things and big data analytics to function effectively. Cloud computing also helps in automizing, integrating and facilitating management and administration of resources and client-based server systems (Foster et al., 2018). Some notable examples of cloud systems are Microsoft OneDrive, Google Drive and BlueCloud (Haug et al., 2016).

(iv) *Autonomous Industrial Robots*: Robots today find major applications in manufacturing to solve complex tasks which cannot be solved by humans. But autonomous robots create a human–robot interface to perform tasks at workstations. These autonomous robots are fed with the necessary information by any operator or cloud-based control system and so they can be remotely controlled by humans (Geissbauer et al., 2014). Several new technologies have recently been developed to complete sensitive tasks and improve the ability of the robots to learn, check and optimise tasks with the help of cloud systems (Bahrin et al., 2016).

(v) *3D Printing/Additive Manufacturing*: 3D printing refers to the customised production of goods by the process of building an object by depositing material in multiple layers based on the 3D model data provided (Rüßmann et al., 2015). This technology is most commonly used for prototyping or to produce goods in small batches by gaining the advantages of avoiding overproduction and minimising inventory (Conner et al., 2014). Though 3D printing technology is yet to reach its full potential, it provides Industry 4.0 with the technological infrastructure to produce customised products (Frazier, 2014).

(vi) *Simulation*: Simulation software serves as the digital twin to the physical world of machines which allows virtually testing them and optimising the process to increase quality (Weyer et al., 2016). Recently firms have begun to take extensive advantage of advances in 3D simulation software in planning plant operations of products and production processes. These digital tools provide the capability of configuring shop floor management systems and also re-configuring existing ones for increased effectiveness. Simulation offers adjustments to the individual parameters and adjustments of complex systems and accurately estimates the system outputs (Brettel et al., 2014). Thus, simulation can serve as a dynamic investigation tool for strategic planning with real-time data acquisition.

(vii) *Augmented Reality*: This is the interactive technology that enables the users to interact with a virtual world built on the basis of real-life surroundings. This technology enables human–machine interactions and remote controlling of factory maintenance and inspection tasks (Fraga-Lamas et al., 2018). Augmented reality-based service and support systems provide virtual training to learn to interact with machines (Longo et al., 2017). These systems and technologies are still in their infancy, and in the future, there is huge potential for much broader use of augmented reality to provide real-time information on the factory operations which can enhance accurate decision-making and help in developing efficient work procedures (Masoni et al., 2017).

(viii) *Enhanced Cybersecurity*: Many companies are still hesitant to put their company data online because they do not trust the existing cyber-systems. The fact that the cyber-systems are hackable and allow misuse of critical information have them on the back foot, still relying on the unconnected and closed production management systems (Kobara, 2016). Increased connectivity and advanced communication systems in Industry 4.0 must be capable of protecting industrial systems and manufacturing lines from any threats and terror incidents (Cho & Woo, 2017). Cybersecurity measures increase confidentiality, integrity and availability along with information privacy (Roy et al., 2016). Hence cybersecurity is considered an important issue and therefore preventive solutions and defensive systems are a major part of Industry 4.0.

(ix) *Horizontal and Vertical IT Integration*: Most IT systems in existence today are fully integrated and Industry 4.0 demands the close linking of OEMs, suppliers and customers on the one hand, and on the other, linkage from shop floor operations to the enterprise operations (Deloitte, 2015). Horizontal integration refers to the integration of various IT systems used in different stages of manufacturing and business planning involved in exchange of materials, energy and information to provide end-to-end solutions and vertical integration of IT systems at different hierarchical levels, from actuators and sensors to corporate planning (Wang et al., 2016). This type of integration enables Industry 4.0 companies to manage complex tasks of exchanging product and production data among multiple partners dynamically.

1.1.4 BENEFITS OF INDUSTRY 4.0

Industry 4.0 has a huge potential to profitably satisfy individual customer requirements with its inherent capability to address last-minute production changes or any unforeseen disruptions (Yin et al., 2018). The end-to-end transparent manufacturing process provides the option of enhanced value creation and facilitates optimised decision-making (Kagermann et al., 2013). But to successfully implement Industry 4.0, the industrial policy decisions, R & D decisions and even the operational decisions need to work alongside and support each other. In the following subsections, we will discuss the potential of Industry 4.0 in detail.

> *Satisfying Individual Customer Requirements*: efficient mass production systems that evolved as a result of previous industrial revolutions failed to address the customised requirements of individual customers (Schwab, 2017). Though concepts like flexible manufacturing systems and the delayed customisation of products tried to address this issue, they were not effective enough to handle last-minute changes in production or delivery. Industry 4.0 with a smart factory enables firms to make a profit even in cases of low production volumes, even, say, in the case of single batch production (European Commission, 2013).

> *Enhanced Flexibility and Adaptability*: In today's dynamic market environment, industries are forced to handle last minute changes in manufacturing lines with greater agility, handle any shortcomings and disruptions from their suppliers and still be a profitable business (Bag et al., 2018). Usually these problems were difficult to handle due to the absence of networking systems which allow dynamic transfer of information. Cyber-physical systems, the heart of Industry 4.0, enable agile manufacturing and agile supply chains to achieve increased output in a short duration by accommodating temporary shortages (Ardanza et al., 2019). On the other hand, adaptability is required to react to changes beyond the predetermined scale of actions and to proactively react to market changes. The reconfigurable manufacturing systems in Industry 4.0 can enable industries to change and modify their processes and systems by simple rebuilding (Bahrin et al., 2016).

> *Resource and Energy Efficiency*: With a growing population and increased utility of resources, our planet faces serious a resource crisis and need for new circular business models (Schwab, 2017). Consumption of excessive

raw materials and energy also leads to serious environmental threats like climate change. Industry 4.0 provides the necessary trade-offs between the excess resource utilisation and potential savings by means of accurate calculations offered by smart factories (Yin et al., 2018). The cyber-physical systems allow tracking of the entire value chain to optimise the manufacturing process in terms of resources utilisation, energy consumption and even reduced emissions. Basically, traditional machines can be transformed into smart machines which are sophisticated and capable of sharing live information continuously to allow better coordination of work, resulting in efficient optimisation of production lines leading to quality production and better resource utilisation (Wang et al., 2016).

Better Work–Life Balance: Smart factories and smart assistance systems enable the firms to overcome shortages of skilled labour and eliminate the need for human involvement in difficult tasks (Wang et al., 2016). This in turn benefits the existing employees by allowing them to focus on innovative and value-added tasks rather than wasting their valuable time and energy on regular routines (Weyer et al., 2016). Digitalisation has enabled decentralisation and made organisations flexible enough promoting a better work–life balance by enabling the employees to concentrate well in their private lives and still be able to work and continue their professional development more efficiently and effectively (Schwab, 2017). There is also a risk that companies with effective cyber-physical systems will reduce the need for human workforce, but these companies will be able to recruit the best employees. This will result in a competitive but high wage economy.

Research and Technological Advances: Adopting Industry 4.0 technologies will push firms to concentrate more on cybersecurity and providing technical education to their employees (Weyer et al., 2016). Though we have stepped into the fourth industrial revolution, connecting the products with the internet, real-time control technologies, connecting business partners, smart manufacturing process and 3-D printing are still in their infancy and the scope of research in these areas has grown considerably (Xu et al., 2018). Manufacturers will still need to ensure that the appropriate cybersecurity is in place to physically secure the firms' data and also will have liability if a breach occurs (Kobara, 2016). Basically, the firms will encourage advanced research and technologies to develop contingency and response plans to counter any cyber-attacks. Thus, Industry 4.0 has immersed the firms in research and innovation to explore and understand new technologies and their potential impacts.

1.1.5 Conclusion

This introductory chapter summarises the evolution of Industry 4.0 by highlighting the key historical events triggering the previous revolutions. The key developments and fundamental differences between the four revolutions are discussed based on how transformation of the industries occurred. A definitions-based explanation of Industry 4.0 is presented with the help of literature. Further, the key components of Industry 4.0 and the benefits associated with its successful implementation are emphasised. Though the concept of Industry 4.0 has been a buzzword in both industry and academia for nearly a decade, its implementation is found to be in its infancy and industries struggle to successfully adopt it, because most of them are not ready for it and also due to the lack of proper implementation framework. This calls for future research on implementation framework models and advanced research on supporting technologies.

REFERENCES

Ardanza, Aitor, Aitor Moreno, Álvaro Segura, Mikel de la Cruz, and Daniel Aguinaga. "Sustainable and flexible industrial human machine interfaces to support adaptable applications in the Industry 4.0 paradigm." *International Journal of Production Research* 57, no. 12(2019): 1–15.

Bag, Surajit, Arnesh Telukdarie, J. H. C. Pretorius, and Shivam Gupta. "Industry 4.0 and supply chain sustainability: Framework and future research directions." [ahead of print] *Benchmarking: An International Journal* (2018). doi: 10.1108/BIJ-03-2018-0056.

Bagheri, Behrad, Shanhu Yang, Hung-An Kao, and Jay Lee. "Cyber-physical systems architecture for self-aware machines in industry 4.0 environment." *IFAC-PapersOnLine* 48, no. 3 (2015): 1622–1627.

Bahrin, Mohd Aiman Kamarul, Mohd Fauzi Othman, NH Nor Azli, and Muhamad Farihin Talib. "Industry 4.0: A review on industrial automation and robotic." *Jurnal Teknologi* 78, no. 6–13 (2016): 137–143.

Brettel, Malte, Niklas Friederichsen, Michael Keller, and Marius Rosenberg. "How virtualization, decentralization and network building change the manufacturing landscape: An industry 4.0 perspective." *International Journal of Mechanical, Industrial Science and Engineering* 8, no. 1 (2014): 37–44.

Brynjolfsson, Erik, and Andrew McAfee. *The Second Machine Age: Work, Progress, and Prosperity in a Time of Brilliant Technologies.* New York: WW Norton & Company, 2014.

Cho, Hyo Sung, and Tae Ho Woo. "Cyber security in nuclear industry–Analytic study from the terror incident in nuclear power plants (NPPs)." *Annals of Nuclear Energy* 99 (2017): 47–53.

Conner, Brett P., Guha P. Manogharan, Ashley N. Martof, Lauren M. Rodomsky, Caitlyn M. Rodomsky, Dakesha C. Jordan, and James W. Limperos. "Making sense of 3-D printing: Creating a map of additive manufacturing products and services." *Additive Manufacturing* 1 (2014): 64–76.

Deloitte, A. G. "Industry 4.0 challenges and solutions for the digital transformation and use of exponential technologies." *McKinsey Global Institute* 13 (2015): 1–16.

Drath, R. and A. Horch. "Industrie 4.0: Hit or hype?" [industry forum]. *IEEE Industrial Electronics Magazine* 8, no. 2 (2014): 56–58.

European Factories of the Future Research Association. "Factories of the future: Multi-annual roadmap for the contractual PPP under Horizon 2020." Brussels, Belgium: Publications office of the European Union, 2013.

Foster, D., L. White, J. Adams, D. C. Erdil, H. Hyman, S. Kurkovsky, M. Sakr, and L. Stott. "Cloud computing: Developing contemporary computer science curriculum for a cloud-first future." In *Proceedings Companion of the 23rd Annual ACM Conference on Innovation and Technology in Computer Science Education*, Larnaca, Cyprus, (July 2018): pp. 130–147.

Fraga-Lamas, P., T. M. Fernández-Caramés, O. Blanco-Novoa, and M. A. Vilar-Montesinos. "A review on industrial augmented reality systems for the industry 4.0 shipyard." *IEEE Access* 6 (2018): 13358–13375.

Frazier, William E. "Metal additive manufacturing: A review." *Journal of Materials Engineering and Performance* 23, no. 6 (2014): 1917–1928.

Freeman, Christopher, and L. L. G. Soete. *The Economics of Industrial Innovation*, 354–355. Technical Change and Economic Theory, London: Pinter, 1997.

Geissbauer, R., S. Schrauf, and V. Koch. "Industry 4.0–opportunities and challenges of the industrial internet assessment." PricewaterhouseCoopers, 2014. https://www.pwc.nl/en/assets/documents/pwc-industrie-4–0.pdf.

Gorecky, D., M. Schmitt, M. Loskyll, and D. Zühlke. "Human-machine-interaction in the industry 4.0 era." In *2014 12th IEEE International Conference on Industrial Informatics (INDIN)*, IEEE, (July 2014): 289–294.

Haug, Katharina Candel, Tobias Kretschmer, and Thomas Strobel. "Cloud adaptiveness within industry sectors–Measurement and observations." *Telecommunications Policy* 40, no. 4 (2016): 291–306.

Hellinger, Ariane, and Heinrich Seeger. "Cyber-physical systems. Driving force for innovation in mobility, health, energy and production." *Acatech Position Paper, National Academy of Science and Engineering* 1, no. 2 (2011).

Hofmann, E. and M. Rüsch. "Industry 4.0 and the current status as well as future prospects on logistics." *Computers in Industry* 89 (2017): 23–34.

Kagermann, Henning, Johannes Helbig, Ariane Hellinger, and Wolfgang Wahlster. "Recommendations for implementing the strategic initiative INDUSTRIE 4.0: Securing the future of German manufacturing industry; final report of the Industrie 4.0 Working Group." *Forschungsunion*, 2013.

Klingenberg, C., and José Antônio do Vale Antunes Jr. "Industry 4.0: What makes it revolution." In *Predavanje na konferenci 24th International EurOMA conference Edinburgh: Inspiring Operations Management*, Edinburgh, vol. 1, no. 5 (2017).

Kobara, Kazukuni. "Cyber physical security for industrial control systems and IOT." *IEICE TRANSACTIONS on Information and Systems* 99, no. 4 (2016): 787–795.

Landes, David S. *The Unbound Prometheus: Technological Change and Industrial Development in Western Europe from 1750 to the Present*. United Kingdom: Cambridge University Press, 2003.

Lee, J., B. Bagheri, and H. A. Kao. "A cyber-physical systems architecture for industry 4.0-based manufacturing systems." *Manufacturing Letters* 3 (2015): 18–23.

Lee, J., H. A. Kao, and S. Yang. "Service innovation and smart analytics for industry 4.0 and big data environment." *Procedia Cirp* 16, no. 1 (2014): 3–8.

Longo, Francesco, Letizia Nicoletti, and Antonio Padovano. "Smart operators in industry 4.0: A human-centered approach to enhance operators' capabilities and competencies within the new smart factory context." *Computers & Industrial Engineering* 113 (2017): 144–159.

Loshkareva, E., P. Luksha, I. Ninenko, I. Smagin, and D. Sudakov. "Skills of the future. Which knowledge and skills are necessary in the new complex world." 2015. http://worldskills.ru/media-czentr/dokladyi-i-issledovaniya.html (accessed October 21, 2017).

Masoni, Riccardo, Francesco Ferrise, Monica Bordegoni, Michele Gattullo, Antonio E. Uva, Michele Fiorentino, Ernesto Carrabba, and Michele Di Donato. "Supporting remote maintenance in industry 4.0 through augmented reality." *Procedia Manufacturing* 11 (2017): 1296–1302.

Rahman, H. and R. Rahmani. "Enabling distributed intelligence assisted future internet of things controller (FITC)." *Applied Computing and Informatics* 14, no. 1 (2018): 73–87.

Rahman, M. N. Abd, B. Medjahed, E. Orady, M. R. Muhamad, R. Abdullah, and A. S. M. Jaya. "A review of cloud manufacturing: Issues and opportunities." *Journal of Advanced Manufacturing Technology (JAMT)* 12, no. 1 (2018): 61–76.

Rifkin, Jeremy. "How the third industrial revolution will create a green economy." *New Perspectives Quarterly* 33, no. 1 (2016): 6–10.

Roy, Rajkumar, Rainer Stark, Kirsten Tracht, Shozo Takata, and Masahiko Mori. "Continuous maintenance and the future–Foundations and technological challenges." *CIRP Annals* 65, no. 2 (2016): 667–688.

Rüßmann, M., M. Lorenz, P. Gerbert, M. Waldner, J. Justus, P. Engel, and M. Harnisch. "Industry 4.0: The future of productivity and growth in manufacturing industries." *Boston Consulting Group* 9, no. 1 (2015): 54–89.

Schwab, K. *The Fourth Industrial Revolution.* Currency, 2017.

Seele, P. "Predictive sustainability control: A review assessing the potential to transfer big data driven 'predictive policing' to corporate sustainability management." *Journal of Cleaner Production* 153 (2017): 673–686.

Smit, Jan, Stephan Kreutzer, Carolin Moeller, and Malin Carlberg. "Industry 4.0: Study." *European Parliament*, 2016.

Sommer, Lutz. "Industrial revolution-industry 4.0: Are German manufacturing SMEs the first victims of this revolution?" *Journal of Industrial Engineering and Management* 8, no. 5 (2015): 1512–1532.

Wang, Shiyong, Jiafu Wan, Daqiang Zhang, Di Li, and Chunhua Zhang. "Towards smart factory for industry 4.0: A self-organized multi-agent system with big data based feedback and coordination." *Computer Networks* 101 (2016): 158–168.

Weyer, Stephan, Torben Meyer, Moritz Ohmer, Dominic Gorecky, and Detlef Zühlke. "Future modeling and simulation of CPS-based factories: An example from the automotive industry." *IFAC-PapersOnLine* 49, no. 31 (2016): 97–102.

Xu, Lu, Seong-Young Kim, Jie Xiong, Jie Yan, and Han Huang. "Playing catch-up: How less developed nations can jump-start technology innovation." *Journal of Business Strategy* 41, no. 2 (2018): 49–57.

Yin, S. and O. Kaynak. "Big data for modern industry: challenges and trends" [point of view]. *Proceedings of the IEEE* 103, no. 2 (2015): 143–146.

Yin, Yong, Kathryn E. Stecke, and Dongni Li. "The evolution of production systems from Industry 2.0 through Industry 4.0." *International Journal of Production Research* 56, no. 1–2 (2018): 848–861.

1.2 CHARACTERISTICS AND DESIGN PRINCIPLES OF INDUSTRY 4.0

Prateek Saxena and Astha Vijaivargia

1.2.1 INTRODUCTION

The manufacturing industry is going through a recently developed industrial revolution referred to as Industry 4.0. The aim of this ongoing revolution is to optimise the manufacturing process, products, resources, maintenance and recycling through the integration of the Internet of Things (IoT) and cyber-physical systems (CPS). Manufacturing of products has changed from digital to intelligent with the fast growth of information technology, electric and electronic technology and advanced manufacturing technology. Virtual reality technology based on CPS has become significant in the new era. Due to the unique challenges in manufacturing processes and techniques, intelligent manufacturing has evolved in recent years. The manufacturing industry makes products that are highly automatised and mechanised. The inventions of new technologies have led to industrial revolution, ever since the beginning of industrialisation took place [1]. The first industrial revolution took place in

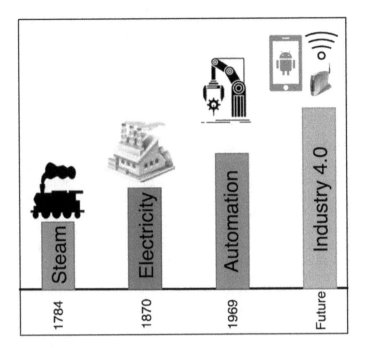

FIGURE 1.2   Four generations of the industrial revolution.

1784. Goods were manufactured using water and steam-powered machines. The subsequent revolutions have resulted in manufacturing by electrical and digital automated production (second and third revolutions). Industry 4.0 is the fourth industrial revolution (shown in Figure 1.2). It is the new stage of industry that can control the entire chain of the product life-cycle. Industry 4.0 is a combination of adaptation and application of digital technology, cyber-systems and the big data approach to transforming the production to the next level. The goal of Industry 4.0 is to attain higher efficiency and productivity, followed by a bigger size of automation and optimisation [2]. The fundamental framework of the continuing fourth industrial revolution is the integration of digitalisation and real-time orientation of various aspects of the manufacturing system.

1.2.2   CHARACTERISTICS OF INDUSTRY 4.0

The five important characteristics of Industry 4.0 are outlined and discussed below.

1.2.2.1   Horizontal and Vertical Integration

Industry 4.0 is mostly defined by three ways of integration: (1) horizontal integration across the entire network, (2) vertical integration and (3) end-to-end integration across the whole life-cycle of a product [3]. Horizontal integration describes 'smart cross-linking of operations performed inside the organization and digitalization of value-added modules throughout the value chain of a life cycle of a product and between life cycles of next product'. End-to-end engineering is across the entire product life-cycle. It describes the smart cross-linking and digitalisation throughout all stages of the life-cycle of a product, beginning from the purchasing of raw materials, to the use of the product and the end life of the product. Vertical integration describes 'the smart cross-linking and digitalization of value creation modules from manufacturing points through manufacturing cells, lines, and factories within the different collection and different levels' [4]. It also combines other activities such as sales and marketing or technology development.

1.2.2.2 Demand and Marketing

Data collection and integration helps in identifying consumer behaviour and developing decision-making models to gain predictive insights. Forecasting, measurement and testing are three goals for analysing the demand. Forecasting understands the results of profit from marketing masterplans and is also vital to the company for forecasting future sales and inventory planning. Hence the demand system is necessary for the decision-making system in the company. A descriptive model is used if the goal of the analysis is to forecast the demand.

Measurement is the second aim of demand analysis. Choosing the correct demand model depends upon what is to be calculated or measured. For example, an analyst cannot measure consumer welfare or risk factors without knowing a customer's utility. An analyst also wanted to know about a consumer's beliefs or collection of inventory based on consumer demand and behaviour. If these are the goals, the structural model is used, which is filled with information on consumer preferences and makes itself clear about the consumer's information sets, trusts and choices. If the goal is to measure the casual effect, then the correct model is that which uses the minimum form.

Testing is the third goal of demand analysis. Testing involves the comparison of at least two theories of the process for data-generation and then decides which theory is right for the demand analysis. The features used in the model for testing demand-side effects are the same as the model used for calculating casual effects of absorbing small forms and incorporating data that can separate the impact of testing [5].

1.2.2.3 Digital Supply Chain and Production

It is the supply chain in manufacturing that is mainly impacted by the introduction of Industry 4.0. All the steps starting from the shipping of the order to the end life of the product are now transparent hence the partnership between suppliers, manufacturers and customers become crucial. The whole supply chain management works by digitalisation and automation of processes. It is necessary to discover the impact of Industry 4.0 on the supply chain so that threats which may arise from the new technologies can easily be identifiable.

The IoT is made up of sharp and intelligent machines, advanced predictive analytics and the partnership of human and machines which improve productivity, reliability and efficiency of the industry. It provides specific sensing/actuating ability of data and information and, therefore, quickly transmits it to the smart devices so that the remote operations of the manufacturing system are greatly facilitated. Analysing, processing and exchanging information between people and machines can be done by the use of smart sensors and actuators. Some of the sensor-based applications can also be used for maintenance of machines as they detect wear and tear of equipment, and also keep a check on inventory levels. IoT technology can thus be utilised for gathering, examining and organising data [6].

1.2.2.4   Digital Products and Services

Consumers' daily products have been changed by recent innovations in microchips, sensor technology and semantic technology. These technologies allow everyday products to be filled with intelligence, sensing and communication abilities. Such a product is referred to as a 'smart product'. Smart products possess unique properties like being context-aware, pro-active and self-organised. They are capable of doing computations, interacting and communicating with their environment and storing data. They not only know their process steps, which they have passed through but they also know their upcoming steps. These steps include not only processing steps that are to be carried out on incomplete goods, but also the future maintenance operations. Smart services are the services provided through smart technologies, and are characterised as easy-to-use, user-friendly and creative services.

All smart products comprise three prime elements: physical parts; smart components (such as sensors); and components for connectivity (like ports and antennae). Smart products have inbuilt smartness and intelligence compared to traditional products. A smart product uses a different combination of technologies, as it is basically information technology (IT) embedded into a tangible product. Some of the examples of smart products and technologies are global sensing technologies such as GPS, local optical sensing technologies like bar codes and QR codes, short-range sensing technologies such as RFIDs, etc.

1.2.2.5   Digital Customer Experience

The rapid growth of smart technologies (e.g., smartphones, tablets, wearables), which were once used by youngsters, are now widely accepted by almost all parts of society because now customers intend to explore the potential of these devices fully. The retailers improve sales by using sensors that catch easy information, such as how many customers have entered the store or their behavioural data. Retailers need to understand customer's demands and their behavioural activities. The relationship between customers and retailers can be increased emotionally by easy and interactive communication and by enjoyable devices. So, the use of smart technology helps customers and retailers to reach mutual goals by achieving better customer experiences [7].

Globalisation, over the past decades, has completely transformed business culture. Geographical boundaries are no longer the boundaries anymore; hence, organisations have a bigger market to chase, though with increased competition. This leads to the launching of new innovative smart products and improving the value of existing products. Appraisal of existing products can be improved by adding new features, driving down cost, minimising defects and therefore improving the overall customer experience.

Customer experience can be improved by smart retail technologies by giving better and individualised services. When customers have to deal with technologically owned services, the concern is how they adopt it and their psychological reactions towards services. Satisfaction and reduction in risk towards smart retail technologies are directly enhanced by smart customer experience. Customers' well-being, their shopping interests and their intentions are increased by customer satisfaction [8].

1.2.3 DESIGN PRINCIPLES

Interoperability, virtualisation, decentralisation, real-time capability, service-orientation and modularity are the six design principles of Industry 4.0. Each of the principles is discussed in the sections below.

1.2.3.1 Interoperability

It is defined as 'the ability to execute the same functions in the manufacturing of a product, even after the interchanging of machines and equipment takes place'. It gives a supportive surrounding in the manufacturing system by extending multiple networks [9]. It is the ability of two systems sharing knowledge and data and exchange information. Interoperability allows sharing of parts of software, business operations and application results throughout the process in the manufacturing system.

The structure of interoperability includes four steps: (1) operational, (2) systematical, (3) technical and (4) semantic interoperability. Operational interoperability shows the general structure of ideas, quality, languages and relations between the cyber-physical system and Industry 4.0. Systematical interoperability identifies the instruction and principles of methodologies, prototypes and standards. Joining of tools and platforms for technological development, information technology systems and related software comes under technical interoperability. Semantic interoperability makes sure that data is exchanged among various categories of people in various levels of the institution; also, malicious packages of applications are removed from the system. Industry 4.0 and CPS become more productive and cost-saving through these four levels of operation [10].

1.2.3.2 Virtualisation

Virtualisation refers to *creating a virtual copy of real-time data*. Virtualisation is used for establishing connections between machines and timely monitoring of the operations. The models used are connected to the sensor data. Virtualisation helps in notifying workers of failures of the manufacturing system so that safety precautions can be taken on time [9]. In a smart factory, the CPS create a virtual copy of the physical world, keeps a record of physical processes and make a decision on their own for the entire process.

A special form of a cyber-physical system is a virtual engineering object (VEO) or virtual engineering process (VEP). CPS merges the physical and digital worlds by creating networks globally in the industry that includes their machinery, production facilities and storage system. VEO is the actual representation of an object which through the experience captures, adds, stores, improves and shares knowledge. VEO can contain the ideas and experience of each crucial detail of an engineering object. VEO can store information which is used to make decisions so that operations can be performed in a better way. It is also useful in the areas of reliability, serviceability and maintainability of the product. VEP represents knowledge of all the operations performed and their sequences in the manufacturing process of a product and types of resources required in manufacturing. VEP determines the sequences for selected manufacturing operations. It also helps in selecting manufacturing resources to 'convert' a design model into a finished product economically and competitively.

To accomplish CPS for Industry 4.0, the main aspect is virtual simulation of goods and processes. In VEO/VEP all the crucial information of process planning is covered by modelling and simulation of manufacturing goods and processes. VEO/ VEP manages with self-organising manufacturing and control planning, and makes a strong bond between product life-cycle management, industrial automation and semantic technologies. Therefore, VEO/VEP is a special structure of CPS [11].

1.2.3.3   Decentralisation

Decentralisation is defined as the *capability of components, machines and working staffs to make their own decisions instead of relying upon the centralised computing decision unit.* For fulfilling the needs of highly customisable products, decentralisation is a perfect tool for manufacturing such products [9]. Decentralised systems have a high capability to manage complex surroundings and customised products. It manages complexity and changes the current centralised production control. In decentralisation, some parts of the manufacturing system, such as machines, products and other elements, make their own decisions without the help of any higher control section. To achieve this, a single element should be able to process data, make confident decisions and perform them. However, to make decisions, all the elements should have access to the information or data stored in the production control unit. The decentralisation method offers small decision-making procedures, since the elements can make decisions on their own and do not depend on the higher control section. This is because decision-making requires less calculation, as there is a limited source of elements decisions. Such an approach is fit for an environment where changes occur frequently, and corrections are made in little time. Due to this approach, the delivery of the products can be made promptly [12].

1.2.3.4   Real-Time Capability

Since the internet is overgrown nowadays, a large amount of data is generated and gathered on a daily basis. Therefore, no standard tools can process and analyse this big data. The rapidly growing database is quickly managed by big data. Or we can say 'big data' refers to the big database which an ideal software tool was not able to capture, store, manage and analyse. This technology analyses and separates the more critical data from the less critical data and transfers effective knowledge to fulfil business activities. With this technology, even the data which is collected in various forms can be collected and analysed to give a clear picture of the situation [13]. Due to big data technologies, the real-time capability of Industry 4.0 reaches a higher level. The collection of big data about the machines, equipment, goods and consumers' data from the manufacturing system is gathered from social media or directly from the selling end, and the data collected from the suppliers, when examined, changes the way of making decisions and gains value for the company [9].

1.2.3.5   Service-Orientation

When radio frequency identification (RFID) and sensors were added to everyday objects in the late 1990s, it laid down the foundation of the IoT. Some years later, it was seen that the sensors were connecting physical things to the internet, called the

IoT, and it started to grow more widely. At the same time, digitisation of consumer essentials, including fitness, healthcare, vehicles and homes, was creating a consumer Internet of Things. This evolution was due to the rapid changes in consumer demand and their growing expectations from the digital world. This also means that the physical world in which we are living at present is rapidly connecting with the digital world, and networks are not just connected with people, but also to places and to objects. Because of this, the consumer attitude towards everyday objects keeps on changing as commonly used items utilise the internet more efficiently and effectively. For example, when a consumer is outside his home, he/she can now switch on or off his lights, television, air-conditioning, etc. with his smartphone.

The consumer Internet of Things can thus be defined as 'Everyday objects and devices that are collected together in the physical environment which are embedded with programmable sensors and actuators, and can connect with the Internet wirelessly'. These smart products interact with themselves and each other and with humans. By sending and receiving information which is stored and organised in the database through the internet, the smart products can interact with each other and humans. To make an extensive network of smart products, IoT is used. IoT also expands the interaction between people and smart products, which is in contrast with the use of the conventional internet, which aims to connect people and exchange information. The machines and objects are embedded with sensors and actuators which allows them to share information over the internet [14].

1.2.3.6   Modularity

Modularity is defined as 'the ability of the modular systems to modify according to the requirement by restoring or elaborating individual modules in ways which are more usable'. Modularity can adapt to situations when there is a need to change the product's production needs. It also provides the possibility of simulating various manufacturing sequences such as product design, production planning, production process and production engineering and interconnect them to offer interchangeability [9]. In smart manufacturing, all the machines, sensors, actuators, people and devices are connected over the IoT and Internet of People (IoP), and together, they form the Internet of Everything (IoE). Wireless communication technologies play an essential role in this interaction. Through the IoE, objects and people who are interconnected can share information and form collaborations to complete common goals. Three types of partnerships are usually associated with the IoE i.e., human–human collaboration, human–machine collaboration and machine–machine collaboration. There is a desire for the development of modular machines for connecting machines, people, devices and sensors. This modularisation further enables smart factories to adapt fluctuation in market demands or customer orders [15].

Another concept within modularity is a reconfigurable manufacturing system. This is a system that is designed to quickly adapt to changes in the market or another requirement similar to this. The system works on the following six capabilities: modularity, integrability, customisation, convertibility, scalability and diagnosability. However, there is no need to achieve all these six capabilities in a typical reconfigurable manufacturing system.

1.2.4 Challenges Involved in Executing Industry 4.0 Framework

Despite the significant advantages of Industry 4.0, there remain specific challenges associated with its actual implementation. Some of the difficulties in implementing the Industry 4.0 framework are as follows:

1. **Data Security:** As the internet is responsible for the increased connectivity among different organisations, the stealing of data becomes very common nowadays. Also, they use a fixed communications protocol, so there remains a requirement for protecting confidential industrial things and their data from cybersecurity threats.
2. **Capital Investment:** For most of the new technology-based industries for production of smart products, investment issues are a big challenge. The execution of all the main parts in Industry 4.0 requires a large amount of capital investment. This will further increase financial risks in the cost of technology development.
3. **Privacy in Data-Sharing:** Data sharing between two companies makes it possible for a third party to have a look over the strategies. So it should be clear from the beginning to whom the generated data belongs and who can use it. In case legal protection is insufficient for some companies, there should be individual contracts. However, having a large number of individual contracts can be expensive for the individual companies.
4. **Unemployment:** As the industries are growing fast with the introduction of advanced machinery and computers, the time will come where all the work is done by the machines and there will be no place for average workers to perform most routine tasks, which will result in unemployment.

1.2.5 Conclusions

The chapter highlights the characteristics and design principles of the fourth industrial revolution called Industry 4.0, which has the potential to develop intelligent, effective and efficient, personalised and customised products at a sensible price. The chapter further discusses demand goals and consumer behaviour in marketing. Warehouse, transport logistics, procurement and fulfilment functions are the technologies implemented in Industry 4.0. They help in understanding opportunities and threats that may be caused by the introduction of new technologies. With the advancements of new technologies, smart products can be used to their full potential to benefit the users. Overall customer experience can be improved by adding new features to the existing products, driving down cost, minimising defects. It is also seen that interoperability shares information and knowledge and exchange data between the two systems. Industry 4.0 and CPS becomes more productive and cost-saving by the structures of interoperability. However, to have a successful execution of the concepts of Industry 4.0, challenges such as data security, capital investment, privacy in data-sharing and unemployment remain to be addressed in establishing smart industries of the future.

REFERENCES

1. H. Lasi, P. Fettke, H.-G. Kemper, T. Feld, and M. Hoffmann, Industry 4.0," *Bus. Inf. Syst. Eng.*, vol. 6, no. 4, pp. 239–242, Aug. 2014.
2. P. Saxena, M. Papanikolaou, E. Pagone, K. Salonitis, and M. R. Jolly, "Digital manufacturing for foundries 4.0," In: Tomsett A. (eds) *Light Metals 2020. The Minerals, Metals & Materials Series.* Cham: Springer. 2020.
3. S. Vaidya, P. Ambad, and S. Bhosle, "Industry 4.0: a glimpse," *Procedia Manuf.*, vol. 20, pp. 233–238, 2018.
4. T. Stock and G. Seliger, "Opportunities of sustainable manufacturing in industry 4.0," *Procedia CIRP*, vol. 40, pp. 536–541, Jan. 2016.
5. P. K. Chintagunta and H. S. Nair, "Structural workshop paper discrete-choice models of consumer demand in marketing," *Mark. Sci.*, vol. 30, no. 6, pp. 977–996, 2011.
6. B. Tjahjono, C. Esplugues, E. Ares, and G. Pelaez, "What does industry 4.0 mean to supply chain?" *Procedia Manuf.*, vol. 13, pp. 1175–1182, 2017.
7. P. Foroudi, S. Gupta, U. Sivarajah, and A. Broderick, "Investigating the effects of smart technology on customer dynamics and customer experience," *Comput. Human Behav.*, vol. 80, pp. 271–282, 2018.
8. S. K. Roy, M. S. Balaji, S. Sadeque, B. Nguyen, and T. C. Melewar, "Constituents and consequences of smart customer experience in retailing," *Technol. Forecast. Soc. Change*, vol. 124, no. 2017, pp. 257–270, 2017.
9. S. S. Kamble, A. Gunasekaran, and S. A. Gawankar, "Sustainable industry 4.0 framework: A systematic literature review identifying the current trends and future perspectives," *Process Saf. Environ. Prot.*, vol. 117, pp. 408–425, 2018.
10. Y. Lu, "Industry 4.0: A survey on technologies, applications and open research issues," *J. Ind. Inf. Integr.*, vol. 6, pp. 1–10, 2017.
11. S. I. Shafiq, C. Sanin, E. Szczerbicki, and C. Toro, "Virtual engineering object/virtual engineering process: A specialized form of cyber physical system for industrie 4.0," *Procedia Comput. Sci.*, vol. 60, no. 1, pp.1146–1155, 2015.
12. H. Meissner, R. Ilsen, and J. C. Aurich, "Analysis of control architectures in the context of industry 4.0," *Procedia CIRP*, vol. 62, pp. 165–169, 2017.
13. K. Witkowski, "Internet of things, big data, industry 4.0: innovative solutions in logistics and supply chains management," *Procedia Eng.*, vol. 182, pp. 763–769, 2017.
14. C. L. Hsu and J. C. C. Lin, "An empirical examination of consumer adoption of Internet of Things services: Network externalities and concern for information privacy perspectives," *Comput. Human Behav.*, vol. 62, pp.516–527, 2016.
15. M. Hermann, T. Pentek, and B. Otto, "Design principles for industrie 4.0 scenarios," *Proc. Annu. Hawaii Int. Conf. Syst. Sci.*, vol. 2016-March, pp.3928–3937, 2016.

2 Sustainable Manufacturing and Industry 4.0

2.1 DESIGN FOR SUSTAINABILITY AND ITS FRAMEWORK

K. Lenin, Abdul Zubar Hameed, and M. Fakkir Mohamed

2.1.1 INTRODUCTION

The development towards Industry 4.0 (I.4.0) has had a considerable impact on the producing business. It has supported the institution of good factories, good merchandise and good facilities conjointly known as the Industrial Internet. In addition, acquired and standard business area units developed around these I.4.0 elements. This evolution carries on the third age that was initiated within the 1970s and supported natural philosophy and data engineering gaining strength in automation in producing [1–7]. Finally, a particular opportunity for property producing plays a necessary role within the world, particularly in European countries. According to Revolutionary Organization 17% of gross domestic product is accounted for by business, which conjointly creates some 32 million jobs accompanying many supporting lines of work within the EU [8]. With the event of those technologies, a replacement idea exists.

I.4.0 was initially implemented in the site of Germany throughout the Hanover Trade Fair event in 2011 that epitomised the start of the fourth age. Therefore, this understanding of I.4.0 cannot claim to be the principal one. Additionally, the commercial enterprise arises from a ranked mapping technology that can supervise individuals to meet commercial enterprise I.4.0 [8]. Compelled by these drivers and ruled by I.4.0 style principles, a variety of potential accomplishments is visualised in Figure 2.1. This enrichment stands for possible processes and attainable eventualities towards utilising I.4.0 directions in corporations, involving only the main features to improve the connection of cooperative schemes. Synchronisation between offers, wage claims, personalisation of products and repair refining, decentralisation and in-depth knowledge exercise is driving operational benefit [9].

Depending on the production practice, this stochastic process exhibits a tendency to analyse step-by-step that might resemble steps to implement I.4.0. High-speed I.N.N. protocols: the I.N.N. supporting system used nowadays can't offer sufficient information for serious communication and transfer of high volumes of information; however, it is superior to the wired network in the producing setting. Producing broad and specific precise knowledge, it is a demand to confirm the prime characteristics

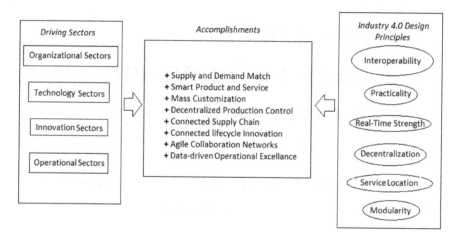

FIGURE 2.1   Potential accomplishments of Industry 4.0.

and principles of the information observed from the producing system. There are many associated notations of knowledge and there is rising demand for analytics [1, 10]. In the systematic approach, to scale back a reduced mathematical statement and finalise an acceptable management model, the system is designed as a self-structured production system. The analysis continues to be created as a self-analysis, occurring for an advanced system, because the world grows fast in conjunction with the Internet of Things (I.o.T.) that galvanised the organisations to initiate an epic journey into I.4.0. I.4.0 will be transformed and explains the current scenario for conversion and automation of production environments [11].

2.1.2   AMONG THE INDUSTRIAL REVOLUTIONS

The first age starts with the mechanism of heat generation around 1795. That brought about the transformation from manual labour to the primary producing process, principally in the fabric business. The second age was the switch over to electrification which led to manufacturing. The third age was distinguished by digitisation accompanied by electronics and automatic control systems (Figure 2.2). In production, this started versatile improvements in manufacturing, whenever a range of products is factory-made on a production line with automatic equipment. These days we have a

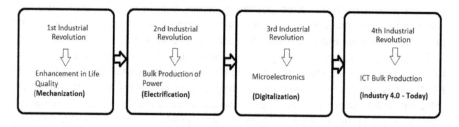

FIGURE 2.2   Through the industrial revolutions.

tendency to transform from all three revolutions to the I.4.0 which is enhanced with knowledge and information communications technology. The I.4.0 scientific idea is crucial in automated cyber-physical systems along with suburbanised management of the latest property [12]. The effects of this latest engineering for industrial manufacturing lines are made with earlier hierarchical automated systems to structured cyber-physical manufacturing lines which permit great manufacturing versatility and flexibility in manufacturing quantity.

The improvement aim is to possess a manufacturing system which will challenge any sort of dynamic industrial enterprise process. So, the system should be distinguished by liability thus it will reply to interruption of assorted root. Options for I.4.0 are parallel, perpendicular and digital desegregation of the whole system. Primary areas are preparing the management of advanced systems, a comprehensive infrastructure, security, privacy, work organization style, legal framework and therefore, the effective use of resources [13].

2.1.2.1   Systematic Changes While Adopting I.4.0

The alterations are centred on the existence interval of product, rather than that specialising in the assembly method [13]. The operational basis of a lean manufacturing industrial enterprise will lead to better manufacturing. In good manufacturing, inspiration can build selections consistent with knowledge, not simply consistent with expertise, the foremost necessary steps to alter what happens within the producing operation, as shown in Figure 2.3 and Figure 2.4.

2.1.2.2   Speculative Stochastic Process of I.4.0

The internet transformation of digital trade continues to be ongoing; however, AI, big data and goods reveal the understanding of the brand new era of the digital age. I.4.0 is on the way, and it can have a crucial influence on the whole transformation of industry as a result. It represents progress on three points:

- Digitalise the manufacturing information network for managing and manufacturing scheduling.

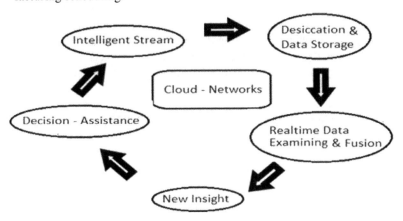

FIGURE 2.3   Concept of a smart factory.

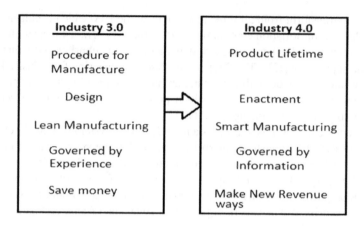

FIGURE 2.4   Vital changes in Industry 4.0.

- Automated networks for information inheritance from the assembly section and virtualisation machines.
- Connect producing sections in a very large supply chain – automated information exchange.

During this period, we tend to use expert automating processes which will require explicit automaticity from the staff. People should still get to use their brains. More money is found from new merchandise, and new solutions for handling figures isn't productive work [14].

2.1.3   APPLYING SUSTAINABILITY TO THE SUPPLY CHAIN

To correlate goods to the method, the three-base goods manufacturing methods are coupled with current method exemplary i.e., efficient, sufficient and consistent. What is more to 'integrate atmospheric thinking into offer chain management' [15], we have a tendency to join the method pattern to the depreciation of raw materials, make waste means by method pattern the input of energy and water, recreative process to revive atmosphere. Decreasing the input related to creating increased use of resources through the transformation and transport of products results in power and having the ability to comment i.e., recycling products into the provision chain correlates to consistent.

2.1.3.1   I.o.T. Empowered Production for Sustainability

I.o.T.-empowered producing relates to a sophisticated element during which typical manufacturing sources are regenerated using good producing technologies that are able to sense, interconnect and take action with one another mechanically and accommodate producing requirements [16]. Inside I.o.T.-empowered producing environments, human-to-human, human-to-machine and machine-to-machine communications are accomplished, for materials are often empowered by means of application of I.o.T. technologies in producing [11, 16]. The I.o.T. is considered to be a contemporary producing idea under I.4.0 and its latest affiliated advances, like the newest information technology infrastructure for knowledge acquisition and sharing, that greatly influences the goods.

2.1.3.2   Robot Interaction for Human Sustainability

With I.4.0, engineering analysis concentrates on a different path to: a) ensure the security of employees, and b) transcend the limitations of the divided space [17]. The core of the robotic work cycles/second development is the desegregation of the dynamic characteristics of each part. Each protected part, entry context, scenario and standing machine uses a method that will activate the protecting mechanisms for safety purposes so the assembly method can proceed while there are no hazards and interruptions, and that can reach the degree of safety that will meet the legal requirements for employee security while working. A supportive human being-machine collaboration [3, 18] is outlined for an enclosed setting within that assembly method, and equipment effectiveness will be improved by means of joining the adaptability of humans and the reliability of machines. Robotic cycles/sec will change human being-machine interactions with their characteristics of dynamic taste designing, active contact shunning and adaption automation management. From those industrial projects, we will develop the main target of exact analysis of human beings – automation cooperation on significant payload robots.

2.1.4   Correlation of I.4.0 and Sustainability

The passage method into property association fitly decided, looked and secured, to make sure the possibility of winning the transmutation travel [19]. The best sensible revolution could also be the combination of region and method style options mentioned at the start. According to constructive of property pack in sharply incorporative. So that area's wide application to the overall results is also concerning for the long run, because the 'minimum' implies it is basic and in context, the particular area needs better choices instead of just 'acceptable' choices. Assignment pursue consultants and votes and also for explicit cases include all thoughtful steps from beginning perception to the ultimate resurrection [19, 20]. That discussed desegregation has continuous application for the whole method. In 2040, the challenge for new techniques may be double that of important manufacturing in 2013 for three items: metallic elements, terbium and atomic number 75 [19, 21]. On account of these metals, the increasing challenge due to the automated technical amendments is considerably above the rise in demand due to the international economic process. During this method, substitutes or correct utilisation and purification would force technological innovation and analysis.

2.1.5   Concluding Remarks

This chapter showed an original results outlook for I.4.0 properties applied to producing and sustaining activities by dissecting their impact on workflows directly connected to the environmental properties. The division of the situation empowered the U.S.A. to take a lot of specific approach requirements that will be smooth-faced, yet the opportunities which arise through the various matters analysed are strained by the I.4.0 components temporal and stage issues. Through the social purpose of read appropriation of I.4.0 by firms, accurate training is followed because the results of the designed and developed tactics may improve the wealth of the whole community and provide the necessary goods. Finally, the development of I.4.0 results in decreased energy use and increased materials and renewable energy.

REFERENCES

1. Stock, T., & Seliger, G. (2016). Opportunities of Sustainable Manufacturing in Industry 4.0. *Procedia CIRP*, 40, 536–541. doi:10.1016/j.procir.2016.01.129.
2. Trstenjak, M., & Cosic, P. (2017). Process Planning in Industry 4.0 Environment. *Procedia Manufacturing*, 11, 1744–1750. doi:10.1016/j.promfg.2017.07.303.
3. Thoben, K., Wiesner, S., & Wuest, T. (2017). "Industrie 4.0" and Smart Manufacturing – A Review of Research Issues and Application Examples. *International Journal of Automation Technology*, 11(1), 4–16. doi:10.20965/ijat.2017.p0004.
4. Mabkhot, M., Al-Ahmari, A., Salah, B., & Alkhalefah, H. (2018). Requirements of the Smart Factory System: A Survey and Perspective. *Machines*, 6(2), 23. doi:10.3390/machines6020023.
5. Foidl, H., & Felderer, M. (2016). Research Challenges of Industry 4.0 for Quality Management. *Innovations in Enterprise Information Systems Management and Engineering Lecture Notes in Business Information Processing*, 121–137. doi:10.1007/978-3-319-32799-0_10.
6. Gibson, R. B. (2009). Beyond The Pillars: Sustainability Assessment as a Framework for Effective Integration of Social, Economic and Ecological Considerations in Significant Decision-Making. *Tools, Techniques and Approaches for Sustainability*, 389–410. doi:10.1142/9789814289696_0018.
7. Piccarozzi, M., Aquilani, B., & Gatti, C. (2018). Industry 4.0 in Management Studies: A Systematic Literature Review. *Sustainability*, 10(10), 3821. doi:10.3390/su10103821.
8. Qin, J., Liu, Y., & Grosvenor, R. (2016). A Categorical Framework of Manufacturing for Industry 4.0 and Beyond. *Procedia CIRP*, 52, 173–178. doi:10.1016/j.procir.2016.08.005.
9. Santos, C., Mehrsai, A., Barros, A., Araújo, M., & Ares, E. (2017). Towards Industry 4.0: An Overview of European strategic roadmaps. *Procedia Manufacturing*, 13, 972–979. doi:10.1016/j.promfg.2017.09.093.
10. Vaidya, S., Ambad, P., & Bhosle, S. (2018). Industry 4.0 – A Glimpse. *Procedia Manufacturing*, 20, 233–238. doi:10.1016/j.promfg.2018.02.034.
11. Arnold, C., Kiel, D., & Voigt, K. (2016). How The Industrial Internet of Things Changes Business Models in Different Manufacturing Industries. *International Journal of Innovation Management*, 20(08), 1640015. doi:10.1142/s1363919616400156.
12. Rojko, A. (2017). Industry 4.0 Concept: Background and Overview. *International Journal of Interactive Mobile Technologies (iJIM)*, 11(5), 77. doi:10.3991/ijim.v11i5.7072.
13. Albers, A., Gladysz, B., Pinner, T., Butenko, V., & Stürmlinger, T. (2016). Procedure for Defining the System of Objectives in the Initial Phase of an Industry 4.0 Project Focusing on Intelligent Quality Control Systems. *Procedia CIRP*, 52, 262–267. doi:10.1016/j.procir.2016.07.067.
14. Roblek, V., Meško, M., & Krapež, A. (2016). A Complex View of Industry 4.0. *SAGE Open*, 6(2), 215824401665398. doi:10.1177/2158244016653987.
15. Man, J. C., & Strandhagen, J. O. (2017). An Industry 4.0 Research Agenda for Sustainable Business Models. *Procedia CIRP*, 63, 721–726. doi:10.1016/j.procir.2017.03.315.

16. Zhong, R. Y., Xu, X., Klotz, E., & Newman, S. T. (2017). Intelligent Manufacturing in the Context of Industry 4.0: A Review. *Engineering*, 3(5), 616–630. doi:10.1016/j. eng.2017.05.015.
17. Vuksanović, D., Ugarak, J., & Korčok, D. (2016). Industry 4.0: The Future Concepts and New Visions of Factory of the Future Development. *Proceedings of the International Scientific Conference - Sinteza 2016*. doi:10.15308/sinteza-2016-293-298.
18. Wang, S., Wan, J., Li, D., & Zhang, C. (2016). Implementing Smart Factory of Industrie 4.0: An Outlook. *International Journal of Distributed Sensor Networks*, 12(1), 3159805. doi:10.1155/2016/3159805.
19. Bonilla, S., Silva, H., Silva, M. T., Gonçalves, R. F., & Sacomano, J. (2018). Industry 4.0 and Sustainability Implications: A Scenario-Based Analysis of the Impacts and Challenges. *Sustainability*, 10(10), 3740. doi:10.3390/su10103740.
20. Müller, J. M., Kiel, D., & Voigt, K. (2018). What Drives the Implementation of Industry 4.0? The Role of Opportunities and Challenges in the Context of Sustainability. *Sustainability*, 10(1), 247. doi:10.3390/su10010247.
21. Beifert, A., Gerlitz, L., & Prause, G. (2018). Industry 4.0 – For Sustainable Development of Lean Manufacturing Companies in the Shipbuilding Sector. *Lecture Notes in Networks and Systems Reliability and Statistics in Transportation and Communication*, 563–573. doi:10.1007/978-3-319-74454-4_54.

2.2 ORIENTATION OF SUSTAINABLE PRODUCT DEVELOPMENT

Abdul Zubar Hameed, K. Lenin, and M. Fakkir Mohamed

2.2.1 INTRODUCTION

Engineering has been enhanced at a fast rate on the enterprise level. Nevertheless, these engineering developments don't seem to have been appropriated to the industrial stages and it is still at the previous stage because the beginning of the third revolution concerned IT performance and support creation. A lot of varied phrases exist for I.4.0 [1]. Varied teams and firms define the term in keeping with their information and view of the discussion [2–13]. In those circumstances, there are interrelating areas like I.o.T., cyber-physical systems (C.P.S.), production systems, medical aid and smart factories [14].

Smart products are products that have the capacity to make estimations, store data, impart information and transfer it to the surroundings. Ranging from timely plans, or a product authorised to spot them via R.F.I.D., the abilities of products to produce information are described. These days, good products do not solely offer their identity, however; they also describe their properties, status and history [14, 15]. Good products are ready to impart information on their life-cycle. They do not just understand the steps of the method as discussed; they are also ready to outline subsequent future steps. These actions embrace not only manufacturing actions still to be carried out on the unfinished product, however; they also cover the forthcoming procedures [15]. Figure 2.5 illustrates the stages of revolution.

As an outcome of the unveiling of steam-based mechanical production, the primary technological revolution began at the end of the 1800s. Electrical power-based manufacturing and section of labour crystal rectifier led to the second technological

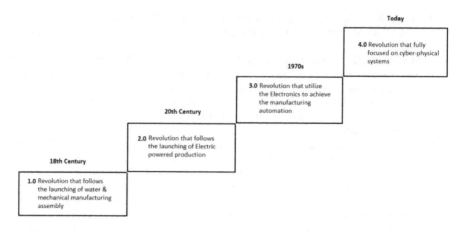

FIGURE 2.5   Stages of revolution.

revolution at the end of the 1900s. From the 1970s on, the utilisation of information technology and advanced natural philosophy for any automated production resulted in the third technological revolution [4, 15]. Thus, the fourth technological revolution is the first revolution that is declared before it proceeds by several firms and analysis institutes (i.e., the common organisation of the Industry 4.0 platform). Industry 4.0 cooperative science laboratories are making an attempt to communicate about the fourth technological revolution so as to give it a longer term.

2.2.2   CYBER-PHYSICAL SYSTEMS

Such systems are composed of manufacturing utilities, storage systems and intelligent machines that initiate actions, exchange information completely and are qualified to control one another, respectively. As a result, firms will connect and consolidate their manufacturing utilities, storage systems and machinery as C.P.S.s within the association of a world wide web [17]. In particular, C.P.S.s are often seen as a brand-new formation of systems that unify the physical world and also the cyber world through group action 'computational and physical capabilities that may act with humans through several new modalities' [14]. To begin with, C.P.S.s adjusted many 'collaborating process entities that square measure in intensive reference to the encircling physical world and its on-going processes' [14]. This may be accomplished by C.P.S.s that are qualified to act and communicate through utilising the knowledge and data from the natural world as well, because the practical world thinks about this adjoining information as an outcome because of the performance of their task [18]. Figure 2.6 explains the composition of the automation pyramid.

2.2.3   INTERNET OF THINGS

At the time of confine, the internet of things (I.o.T.), initially empowered, it absolutely was spoken unambiguously recognisable practical conjoined objects mistreatment radio-frequency identification (R.F.I.D.) engineering. By connecting an R.F.I.D. interpreter to the web, the interpreters will mechanically and unambiguously

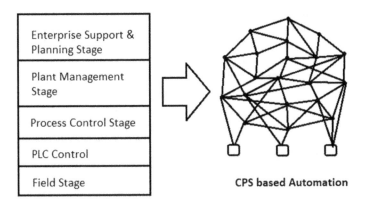

FIGURE 2.6 Composition of the automation pyramid.

establish and trace the target accompanied by the relevant time period. This can be a use for the internet of things (I.o.T.). Consequently, I.o.T. technology can be supported through alternative technologies like sensors, actuators, Global Positioning System (G.P.S.) and mobile devices which are produced via Wi-Fi, Bluetooth, cellular networks or close-to-surface communication. Thence, a newer description of I.o.T. as practised is [19]: a dynamic world wide web substructure with self-confirmation abilities supported by means of commonplace and practical conversation protocols wherever physical and practical 'things' possess individualities, natural dimensions and practical personalities, use acute surfaces and are jointly incorporated into the knowledge web.

2.2.3.1 I.o.T. Employed within Production Management

In I.o.T. exercises in production businesses, production management gathers and handles information on the product description, methods and call over the complete exhibition platform. Distinctive identifiers for products are necessary for production management throughout the limited life-cycle as a result of the product in the P.L.M. area unit disposed not solely at an intra-organisational level, however, together at an inhume-organisational level, in an exceedingly assigned, mobile and cooperative atmosphere [19, 20]. In this case, it is difficult to sustain the mixing of assigned and heterogeneous product information across a completely distinct life-cycle platform with an organised and versatile model. Because of the amount of heterogeneous information and a fast ever-changing atmosphere, the concerned information and data are sophisticated for the exchanging and sharing functions [21]. Figure 2.7 illustrates the I.o.T. of Allied Engineering.

2.2.4 CLOUD COMPUTING

Cloud computing supplies and the disposition producing supplies in the cloud automatic data-processing system primarily see the resources concerned in every manufacturing action throughout the whole life-cycle of a product, as well as material supplies (like production machinery, producing cells, product runs, tools, material, products, software and also abilities because of partitioning a producing drawback

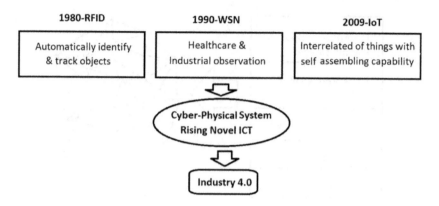

FIGURE 2.7 I.o.T. allied engineering prepared for realisation of I.4.0.

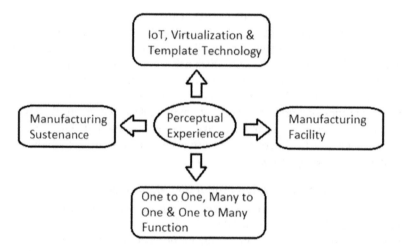

FIGURE 2.8 Connection between production supplies and services in cloud computing.

like human resource, producing data, etc. [19, 21]. There are completely distinct forms of disposition regarding producing supplies in step with different aims, for instance, in step with the two aspects of a replacement yield (i.e., assembly of merchandise and when the merchandise is exhibited), producing supplies are distributed into manufacturing-allied supplies (e.g., raw materials and semiproducts, producing instrumentation, connected style and producing computer software and datapoints regarding the exact data, product formation data and method information) and product-associated supplies (e.g., light and exhausting supplies concerned within the method of transport, services, sustenance, rework, repeat and end life usage). Figure 2.8 explores the connection between production supplies.

2.2.5 CONCLUSION

Finally, understanding the enlightened recognition and approach of producing supplies and abilities is one of the bottlenecks for the imparting of cloud computing,

I.o.T. or cyber-physical systems. While not presenting this downside, the higher executors to produce resources and duties pretend to displace the essential data and information sustained. Industry 4.0 is practical in production to show its importance in the production business. Thence, the main focus is to investigate the orientation that is to be sustained for the product development, whatever the challenges to debate new instructions for I.4.0 that are attainable in the production business for enhancing the I.4.0 through the latest technology. In future, sustainability will play a major role in product development with the help of business intelligence and knowledge management.

REFERENCES

1. Mrugalska, B., & Wyrwicka, M. K. (2017). Towards Lean Production in Industry 4.0. *Procedia Engineering*, 182, 466–473. doi:10.1016/j.proeng.2017.03.135.
2. Barreto, L., Amaral, A., & Pereira, T. (2017). Industry 4.0 Implications in Logistics: An Overview. *Procedia Manufacturing*, 13, 1245–1252. doi:10.1016/j.promfg.2017.09.045.
3. Zezulka, F., Marcon, P., Vesely, I., & Sajdl, O. (2016). Industry 4.0 – An Introduction in the Phenomenon. *IFAC-PapersOnLine*, 49(25), 8–12. doi:10.1016/j.ifacol.2016.12.002.
4. Rojko, A. (2017). Industry 4.0 Concept: Background and Overview. *International Journal of Interactive Mobile Technologies (iJIM)*, 11(5), 77. doi:10.3991/ijim.v11i5.7072.
5. Albers, A., Gladysz, B., Pinner, T., Butenko, V., & Stürmlinger, T. (2016). Procedure for Defining the System of Objectives in the Initial Phase of an Industry 4.0 Project Focusing on Intelligent Quality Control Systems. *Procedia CIRP*, 52, 262–267. doi:10.1016/j.procir.2016.07.067.
6. Wang, S., Wan, J., Li, D., & Zhang, C. (2016). Implementing Smart Factory of Industrie 4.0: An Outlook. *International Journal of Distributed Sensor Networks*, 12(1), 3159805. doi:10.1155/2016/3159805.
7. Khan, A., & Turowski, K. (2016). A Perspective on Industry 4.0: From Challenges to Opportunities in Production Systems. *Proceedings of the International Conference on Internet of Things and Big Data*. doi:10.5220/0005929704410448.
8. Schmidt, R., Möhring, M., Härting, R., Reichstein, C., Neumaier, P., & Jozinović, P. (2015). Industry 4.0 - Potentials for Creating Smart Products: Empirical Research Results. *Business Information Systems Lecture Notes in Business Information Processing*, 16-27. doi:10.1007/978-3-319-19027-3_2.
9. Khan, A., & Turowski, K. (2016). A Survey of Current Challenges in Manufacturing Industry and Preparation for Industry 4.0. *Proceedings of the First International Scientific Conference "Intelligent Information Technologies for Industry" (IITI'16) Advances in Intelligent Systems and Computing*, 15–26. doi:10.1007/978-3-319-33609-1_2.
10. Morrar, R., & Arman, H. (2017). The Fourth Industrial Revolution (Industry 4.0): A Social Innovation Perspective. *Technology Innovation Management Review*, 7(11), 12–20. doi:10.22215/timreview/1117.
11. Prause, G. (2015). Sustainable Business Models and Structures for Industry 4.0. *Journal of Security and Sustainability Issues*, 5(2), 159–169. doi:10.9770/jssi.2015.5.2(3).
12. Wang, C., Bi, Z., & Xu, L. D. (2014). IoT and Cloud Computing in Automation of Assembly Modeling Systems. *IEEE Transactions on Industrial Informatics*, 10(2), 1426–1434. doi:10.1109/tii.2014.2300346.
13. Wortmann, A., Combemale, B., & Barais, O. (2017). A Systematic Mapping Study on Modeling for Industry 4.0. *2017 ACM/IEEE 20th International Conference on Model Driven Engineering Languages and Systems (MODELS)*. doi:10.1109/models.2017.14.

14. Lee, J., Bagheri, B., & Kao, H. (2015). Cyber-Physical Systems Architecture for Industry 4.0-based Manufacturing Systems. *Manufacturing Letters*, 3, 18–23. doi:10.1016/j.mfglet.2014.12.001.

15. Tjahjono, B., Esplugues, C., Ares, E., & Pelaez, G. (2017). What Does Industry 4.0 Mean to Supply Chain? *Procedia Manufacturing*, 13, 1175–1182. doi:10.1016/j.promfg.2017.09.191.

16. Vuksanović, D., Ugarak, J., & Korčok, D. (2016). Industry 4.0: The Future Concepts and New Visions of Factory of the Future Development. *Proceedings of the International Scientific Conference - Sinteza 2016*. doi:10.15308/sinteza-2016-293-298.

17. Arsénio, A., Serra, H., Francisco, R., Nabais, F., Andrade, J., & Serrano, E. (2013). Internet of Intelligent Things: Bringing Artificial Intelligence into Things and Communication Networks. *Inter-cooperative Collective Intelligence: Techniques and Applications Studies in Computational Intelligence*, 1–37. doi:10.1007/978-3-642-35016-0_1.

18. Foidl, H., & Felderer, M. (2016). Research Challenges of Industry 4.0 for Quality Management. *Innovations in Enterprise Information Systems Management and Engineering Lecture Notes in Business Information Processing*, 121–137. doi:10.1007/978-3-319-32799-0_10.

19. Tao, F., Zuo, Y., Xu, L. D., & Zhang, L. (2014). IoT-Based Intelligent Perception and Access of Manufacturing Resource toward Cloud Manufacturing. *IEEE Transactions on Industrial Informatics*, 10(2), 1547–1557. doi:10.1109/tii.2014.2306397.

20. Sanders, A., Elangeswaran, C., & Wulfsberg, J. (2016). Industry 4.0 Implies Lean Manufacturing: Research Activities in Industry 4.0 Function as Enablers for Lean Manufacturing. *Journal of Industrial Engineering and Management*, 9(3), 811. doi:10.3926/jiem.1940.

21. Xu, L. D., Xu, E. L., & Li, L. (2018). Industry 4.0: State of the art and future trends. *International Journal of Production Research*, 56(8), 2941–2962. doi:10.1080/00207543.2018.1444806.

2.3   END OF LIFE DISPOSAL AND SUSTAINABLE INDUSTRIAL WASTE MANAGEMENT IN INDUSTRY 4.0

Alokita Shukla, Rahul Verma, and A. Sofi

2.3.1   INTRODUCTION

For the past few years, every nation has been experiencing rapid population growth. The population of the United States was 296 million in 2005, and it is expected that if it continues to increase at the same rate, then this number may increase to 438 million by 2050. It is estimated that the world population will increase from the current population, which is 7.2 billion, to 9.6 billion by 2050, and further, to 10.9 billion by 2100 (Lutz et al., 1997). Figure 2.9 shows the relation between years and population. The main projection for each country is based on the addition and subtraction of half a child from total fertility rate, in children per women, and the same is provided in the graph (Lee & Tuljapurkar, 1994; Thumerelle, 2001). The abrupt rise in population affects the per capita income and GPD of a country. Some countries practised concurrent growth in population and economics, whereas some experienced only population increases but decreases in economic growth (Bloom & Freeman, 1988). This shows some definite relation between economy and population. Apart from

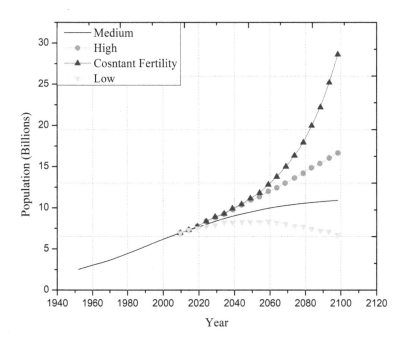

FIGURE 2.9   Relation between population and years (Lutz et al., 1997).

changes in the economy, population rise has some other effects, such as degradation of the environment, increase in pollution and abuse of natural resources. There are a lot of steps which can be taken into consideration to deal with the problems due to the increase in population, and one major step is increasing sustainability. Sustainability refers to minimising the consumption and disposal of something which can cause problems for humans as well as for the environment. The condition of sustainability becomes unbalanced because young and old among the population consume more, but produce less (Bloom et al., 2000; Bloom & Freeman, 1988). With the increase in population, their needs also increase, which becomes a burden on industries. Let us consider the simple example of a nuclear-based power plant. As there is a need for more power because of increased population, the nuclear plant will start to produce more electricity, which can be achieved only by using more fuel, which is uranium. When more power is generated, the amount of water required for cooling of the nuclear reactor increases (Davis et al., 1958). The water used for cooling purposes is extracted from rivers and ponds. After the cooling process, the high-temperature water is transferred back to the river from where it was initially extracted (Kónya & Nagy, 2018). Because of the increase in the water temperature, aquatic life is affected which thus leads to the deaths of certain species of fish and plants. Therefore, it causes water pollution, and purifying the same water for drinking purpose requires more investment due to the presence of impurities and harmful bacteria. These types of investments create economic imbalance for a country.

To overcome such conditions, industrial growth is a primary necessity. In most of the cases, the main focus of an industry is introducing the latest technologies, and not on managing waste. Lack of industrial growth does not only point towards technology, but also depends on awareness when it comes to waste management. A major portion of an economy depends on industries and any development in an industry can alter the economy of a country. Considering the same example of a nuclear power plant, in order to increase the output, some additional technology will be used. As a result, more power will be generated, but along with that, huge investment will also be required for water filtration. Thus, it is necessary to ensure that the effect of industrial development only leads to positive economic growth. Improving such types of technology which lead to negative economic growth is not required. Another important factor which can play a vital role in improvement of the economy is end of life disposal of a product.

2.3.1.1 Effect of End of Life Disposal on Economy

End of life disposal can be defined as using a product only for a particular period of time and then disposing of the product once that period is over. Every product has a specific usable life period and it may vary depending on the type of product. End of life disposal for both biodegradable and non-biodegradable products will be discussed in depth in upcoming sections. The most crucial advantage of end of life disposal is that some of the products can be recycled after the completion of the time period and thus can be used again. The recycling process saves time and is monetarily beneficial. Using products after their specified period can be dangerous for humans as well as for the environment. If edible products are used after their expiry dates, this can cause serious health-related issues (Garcia-Garcia et al., 2016). For vehicles, the average end of life is ten years depending upon the type of vehicle and its usage (Cassells et al., 2005). Continuous use of vehicle even after its prescribed period can results in environmental issues because of excessive smoke from the vehicle exhaust. For the past few years, more preference has been given to the usage of recycled materials and eco-friendly materials instead of sea dumping and incineration (Chan and Tong, 2007). The end of life (E.O.L.) of a product or material should always be specified, and overlooking E.O.L. for a product not only creates environmental issues, but also affects industries. Disposal of products without having an exact E.O.L. (non-biodegradable products) is a difficult task to perform and can bring about unsustainability in waste management. To overcome this issue to the maximum extent, industries are switching towards the use of smart factories, which is also known as Industry 4.0.

2.3.1.2 Brief Introduction of Industry 4.0

Population growth is more than double of its initial growth from 1960 to 2016, which gives rise to three times the purchasing capacity for an individual (Stock & Seliger, 2016; De Man & Friege, 2016). This adversely affects natural resources as well as climate which can lead to social upheaval (Hsiang & Burke, 2014; De Man & Strandhagen, 2017). In the first industrial revolution, steam was used to provide a power supply to mechanical components. The second industrial revolution involved mass production, followed by the third industrial revolution which includes manufacturing with the help of electronics and information technology (Lom et al., 2016; Lucas, 1989). At present, we are in the fourth phase of the industrial revolution, named Industry 4.0, which involves manufacturing processes using the internet of things (I.o.T.). Industry 4.0 was first introduced in 2011 at the Hanover Fair, Germany. The concept behind this was presenting a new economic policy which was based on high-tech strategies (Roblek et al., 2016). Industries coming under this fourth industrial revolution are also known as smart factories. Smart factories include smart machines, smart devices, smart engineering, smart grids and smart manufacturing. The manufacturing processes in smart industries takes place in such a way that it should not have any harmful effect on the environment and should increase sustainability. In this type of industry, the maximum amount of waste is considered for recycling, which leads to sustainability. Let us consider the example of the sugar industry. During the production of sugar, a huge amount of residue is produced which is known as bagasse. Instead of dumping the bagasse, it is used as a fuel to generate steam in boilers. Apart from that, bagasse is also used in the treatment of waste water containing metals (Aly & Daifullah, 1998) and in strengthening of concrete. This shows sustainable waste management and is thus considered as part of Industry 4.0. These industries also accelerate the production rate and therefore can improve the economy of a country. An in-depth study of sustainable waste management in Industry 4.0 will be carried out further in this chapter, along with showing the advantages of Industry 4.0. End of life disposal of various products will also be discussed in the upcoming sections.

2.3.2 END OF LIFE (E.O.L.) DISPOSAL

Nearly 50 years ago during a United Nations Conference on the Human Environment in Stockholm, risks related to 'massive and irreversible harm to the earthly environment on which our life and well-being depends' were thoroughly discussed (Handl, 2012). Since that conference, a number of practices, tools and strategies to regulate the impact of waste on the environment have been proposed. Subsequently, a growing population and an industrial revolution have led to an exponential increase in the production of waste that cannot be recycled (Greyson, 2007). The lack of sustainability in production and consumption patterns have contributed immensely to increasing waste generation at an alarming rate. An early awareness of the importance of controlling waste production and implementation of sustainable management has led to the formulation of many new schemes and techniques. In spite of nascent efforts being made by national, state and municipal organisations and the general public, complete waste elimination is impossible. Thus, to organise the entire waste management system, important steps have to be taken to treat the waste at the end of its serviceable life instead of just trying to prevent this waste generation from its source (Staikos & Rahimifard, 2007).

Proper 'End of Life Disposal' has become the need of the hour due to many factors like excessive dumping of hazardous wastes in landfills which seep into the environment causing water and soil pollution, increased use of incinerators which emit harmful fumes causing air pollution and improper utilisation of recycled products despite their large volumes. For an inexplicably long time, manufacturing companies have completely diverted their efforts to enhancing their marketing, procurement, designing and manufacturing. In this scenario once the manufacturer delivers a product from the factory, they will not have any future contact with the product, apart from its occasional repair or maintenance. However, some recent regulations make companies liable for the proper disposal of the product once its users discard it. This 'take-back' or 'End of Life' disposal has increasingly been imposed as a new requirement on the manufacturers (Toffel & Toffel, 2003; Reimer et al., 2002).

2.3.2.1   End of Life Disposal for Biodegradable Waste

2.3.2.1.1   Textile Products

One of the direct implications of economic development and population growth is an increased demand for textile products, which is expected to increase tremendously in future as well. Textile industries have always been a major contributor towards air, water and soil pollution. Growing wealthy and extravagant lifestyles have led to the emergence of a trend called 'fast-fashion' (Yee et al., 2016) It simply refers to the habit of throwing away apparel after a short usage period which otherwise would have been usable for decades. The main brains behind popularising this trend are the retailers dealing in this fast-fashion. Their simple aim is to increase visits to their stores just to increase their revenue by encouraging the frequent disposal of perfectly useful clothes in the name of fashion upgrading (Bhardwaj & Fairhurst, 2010) To achieve this, these retailers are manufacturing new clothes every week whose prices are kept very low, which in turn further encourages young shoppers to fill their wardrobes with up-to-the-minute fashion garments. Overconsumption of clothes followed by improper disposal of mostly underutilised clothes is becoming the major environmental concern lately (Birtwistle & Moore, 2007).

Treloar et al. (2004) stated that large volumes of textile wastes are filling up land-fills and contributing to pollution by excessive emissions. However, it is not just the disposal which endangers the environment; the production of textile products is an equally serious concern. Petrochemicals are the source for almost 63% of fibres used in textile manufacturing (Lenzing et al., 2017). Producing these fibres causes large emissions of carbon dioxide. The remaining 37% of fibres are derived from cotton, which requires large portions of water during its cultivation. The later stages in textile production often have worse effects than the earlier stages of fibre production. While spinning, knitting and weaving require fossil fuels leading to CO_2 emission, wet treatments like printing and dyeing serve as a major source for the emission of toxic waste (Roos et al., 2015). As per the estimation by Sandin et al. (2015), about 70% of the production must be reduced by 2050 for the textile industry to be sustainable as per the limitations of the Earth stated by Steffan et al. (2015). To achieve such a gigantic feat, the most viable methods are recycling and reuse. The main aim of these methods would be to reduce the manufacturing of new textile fibres which will also help in skipping many stages for production thus reducing the effects of textile production as a whole.

The easiest method for end of life disposal of textiles is to give it away for its reuse after it has been discarded by the initial owner. This textile distribution is often localised without any third person intervention. Such distribution can be as a part of goodwill or the owner may sell the clothes in a flea market or on the internet. However, nowadays large amounts of discarded textiles are being collected by charitable organisations or NGOs. These organisations then segregate the clothes on the basis of type, usability and relevance. Clothes with the best condition or quality are sent for re-sale in second-hand stores at low prices. Others are mostly distributed among the needy. Clothes which are too worn-out for reuse are mostly sent for recycling (Koligkioni et al., 2018).

For sustainable textile waste management in Industry 4.0, recycling of textiles in their end of life stage is the most viable method which needs to be popularised. The efficiency of the textile recycling process is near to 90 per cent, with most of the fibres being recovered (Koligkioni et al., 2018; Wang, 2010). The quantity and quality of the recovered fibres often depend on the method of recycling which may be chemical, mechanical, thermal or an amalgamation of these methods. For instance, if the fabric of the textile waste is in recoverable conditions, it is directly employed to form new products. This is referred to as fabric recycling or as material reuse according to Zamani et al. (2014). When the fibres of the textile waste are recoverable, but not the fabric, the recycling is referred to as fibre recycling. When the degradation of textiles is more intense, oligomer or polymer recycling and further monomer recycling are employed, which involves re-polymerising and re-spinning of yarns. These recycling types cannot be a direct product of one method and thus a combination of treatments is involved (Itorial, 2017). Finally, the recycled product formed after various recycling techniques may be either of lower quality and value or of higher quality and value when compared to the original product. Thus, the former type is considered to be a down-cycled product, while the latter is an up-cycled product. If the original product was a clothing item, its wear and tear during usage and laundry will decrease the lengths of constituent fibres. If fibres are to be recycled, and polymer or monomer recycling is not necessary, down-cycling is the only feasible option. The down-cycled product of this clothing item will often consist of sub-standard blankets, rags or insulations (Schmidt et al., 2016). This is also referred to as open loop cycling, as shown in Figure 2.10. If the recycled product is of comparable value to the original, the process is also called closed loop recycling (Ekvall, 2001). When

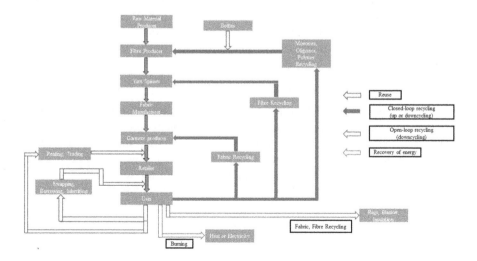

FIGURE 2.10   End-of-life disposal of textile products (Sandin & Peters, 2018).

no recovery even of monomers is possible in textile waste, it is often burnt in incinerators to produce heat and electricity.

2.3.2.1.2   Dairy Products

Dairy waste is mainly comprised of products obtained from agri-based industries like cheese factories and dairy farms. Products from these industries contribute significantly towards the economy worldwide, especially in the Mediterranean region where Protected Designation of Origin (P.D.O.) products, like cheese, are produced. Due to irregularities between the market produce and consumer needs, large quantities of dairy products are dumped in landfills as they approach their end of life state. Disposal of such biodegradable waste is strictly limited and mostly not done in compliance with the 1999/31/EU Landfill Directive (Council Directive 1999/31/EC of 26 April 1999 on the Landfill of Waste). Improper handling without enough treatment of dairy products endangers the environment due to extremely harmful organic by-products.

End of life dairy products contains substantial amounts of carbohydrates, proteins, lipids and mineral salts due to which their chemical and biological oxygen demand is very high thus making them challenging for bioprocessing (Perle et al., 1995). Due to these features, these dairy products are most suitable for anaerobic digestion while concurrently producing bioenergy and thus reducing emission of greenhouse gases (Cantrell et al., 2008). Sustainable dairy waste management can therefore be executed with the help of this technique (Stavropoulos et al., 2016).

Bioenergy production is now paving the way for a sustainable waste management outlook for end of life dairy products. This renewable source of energy fits impeccably into future energy production requirements, especially in terms of economic factors. After the exhaustion of energy based on fossil fuels, gaseous bioenergy fuels like hydrogen can easily substitute for the former while serving as a zero-emission source (Lyberatos, 2010). Hydrogen largely represents the fuel of the future as it is known to have highest energy content per unit weight (143 JG ton^{-1}) as compared to other gaseous fuels, apart from being environmentally friendly even during its production (Boyles, 1984; Ghimire et al., 2015). Thus, hydrogen production from end of life dairy products could prove to be crucial for Industry 4.0.

2.3.2.1.3   Biodegradable Plastics

Plastics or polymers like polypropylene or polythene have a durability of hundreds of years even after they are disposed of. The imperishable nature of plastics proves to be a bane rather than a boon when analysed in the long-run, since various applications of these plastics are normally for shorter periods of time. One such example is polyolefins, which are extensively used in the manufacturing of basic household products like bottles, toys, buckets, plastic bags and many other such items. These items are not expected to have a long shelf-life and are often disposed-of before they actually wear out. As a result, the amount of plastics in dump yards is exponentially increasing every day, further resulting in a gigantic volume shortage problem in the waste management sector (Aminabhavi et al., 2017). Plastics being normally utilised as the most well-known packaging material, are disposed of together with other perishables which make its recycling comparatively difficult and mostly undesirable. Therefore, with the progression of industrialisation and improved concern about the environment, the need for shifting to a biodegradable option from conventional plastics arose. Biodegradable plastics are the ones which when exposed to bioactive conditions are broken down to simpler molecules by either the action of enzymes produced by microorganisms like bacteria, algae and fungi, or by chemical hydrolysis. Biodegradable plastics are manufactured from agricultural or plant wastes. For their end of life disposal, they can easily be sent for composting which will further enhance new plant growth. Thus, biodegradable plastics provide a sustainable alternative which could prove to be revolutionary in waste management. Owing to these advantages, global utilisation of biodegradable plastics in the last decade has increased by approximately 50 million kg. Biodegradable plastics find ample use in packaging, agricultural tools, sanitation products and other day-to-day consumer products. However, its sustainable disposal is still in the rudimentary stage which makes the inexpensive, non-biodegradable plastics an easier choice. Thus, development of an efficient end of life disposal facility for bio-plastics is of the utmost necessity to avoid usage of conventional plastics and decrease the waste plastic accumulation in landfills (Gross & Kalra, 2002). Figure 2.11 shows E.O.L. disposal of biodegradable plastics.

Waste management of biodegradable plastics at their end of life is easier than conventional plastics. Biodegradable plastics, as suggested by the name, are degraded in the presence of microorganisms, light, oxygen and the optimum temperature (Mudgal et al., 2012; Song et al., 2009). The initial step of sustainable end of life disposal of plastics is to segregate them according to their suitability for composting, incineration, landfills or composting. This would require industrial involvement through the introduction of various waste collection schemes. With advancements in technology, segregation of biodegradable plastics is made possible with the help

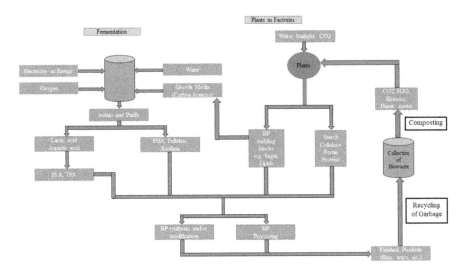

FIGURE 2.11   End of life disposal of biodegradable plastics (Gross & Kalra, 2002).

of infrared spectroscopy or by mechanical separation by their weight and densities (Mudgal et al., 2012).

(i) *Biological Waste Treatments: Composting and Anaerobic Digestion*

Different materials and chemical compositions of biodegradable plastics are suitable for different types of end of life waste management practices. When heterogeneous biodegradable matter is made to undergo an accelerated degradation under controlled conditions, the process is called composting. Composting gives the inherent advantage of converting waste biodegradable plastics to soil enhancement products. Thus for end of life disposal of biodegradable plastics, the most environmentally favourable method is composting if the means to control the bioactive conditions are present (Taylor et al., 2011). On composting biodegradable plastics, nutrient- and carbon-rich compost is produced (Rujnić-Sokele & Pilipović 2017).

(ii) *Recycling (and Reprocessing)*

If strength and durability of the final product are not a major concern, then recycling and reproduction of biodegradable plastics is another significant end of life waste management route. Some bio-plastics like polylactide can easily be recycled for some limited periods without much change in quality and strength. For many applications like packaging, a multi-layered lamination is formed with alternating layers of different biopolymers. This decreases the effectiveness of the process, since not every biodegradable plastic is suitable for recycling (Song et al., 2009). However, the lack of segregated bio-plastic waste is the main restriction which makes recycling a less economical option for biodegradable plastics when compared to conventional plastics (Rujnić-Sokele & Pilipović, 2017).

(iii) *Incineration (and the Other Recovery Options)*

Most biodegradable plastics exhibit a comparable or even higher gross calorific value when compared to coal. When recycling, composting or landfilling is not feasible, then biodegradable plastics can be used for energy recovery applications. The energy recovered from bio-plastics or biodegradable plastics is considered to be a renewable form of energy. Petrochemical carbon used in the incineration process is stated by Song et al. (2009) to be an environmentally better option than direct burning of fuel (Rujnić-Sokele & Pilipović, 2017).

(iv) *Landfill*

Landfill is considered to be the last option of waste management hierarchy and is thus the least preferred too. However, it is still the most widely used end of life disposal method. For instance, a total of around 8 million tonnes of recoverable plastic (30.8% of total) is sent to landfills per year, as stated by the European Union (PlasticsEurope, 2015). This dumping practice is resulting in depletion of appropriate landfill sites across Europe. Moreover, these landfills with biodegradable waste largely produce methane gas which has much higher global warming potential than CO_2. Thus, these landfills contribute to global warming as well. Leaching of harmful chemicals from these landfills is another issue which is thus raising public concern (Mudgal et al., 2012; Song et al., 2009; Rujnić-Sokele & Pilipović, 2017).

2.3.2.2 Footwear Industry

Many industries deal with products which in their end of life stages cannot be classified directly as either biodegradable waste or non-biodegradable waste. They are made up of so many subparts with some parts inherently having biodegradable properties while others have non-biodegradable qualities. One such industry is the footwear industry which requires an enormously diverse range of materials for manufacturing various styles and types of footwear. This diverse material range includes textile materials, rubbers, plastics, leather and other synthetic materials. This wide range requires different end of life disposal routes that can ensure sustainable end of life disposal in Industry 4.0. As stated in earlier works, an average shoe comprises of at least 40 different types of materials (Harvey, 1982). Among them, some are given in Table 2.1.

Since the last few decades, the footwear industry is facing a sudden increase in demand for footwear. Due to fast-changing fashion and volatile consumer habits, the average usage period of a shoe has decreased to less than half. This can also be considered a result of 'fast-fashion', as discussed for textiles. An increase in production by 80% has been observed in the last two decades increasing production to 20 billion pairs every year. In the UK itself, around 330 million pairs of footwear are produced per year, while both the United States and Western Europe yearly produce more than 2 billion pairs of shoes each (World Footwear, 2005; British Footwear Association, 2005). Increased production of footwear is directly responsible for increased production of end of life footwear waste. As per estimations based on earlier works, around 12% of post-consumer footwear is redistributed by N.G.O.s or in garage sales as second-hand footwear, while the remaining 88% is mostly being disposed of in landfills. Large-scale landfill dumping has caused groundwater, soil and surface pollution. These sites thus pose a serious threat to the environment by polluting water sources like lakes and rivers. This has given rise to the need for amendments in environmental legislations to assign responsibility to the producers for proper end of life for footwear products (Abbot & Wilford, 1999).

TABLE 2.1
Material for Footwear Industry (UNIDO, 2000)

Footwear Materials	Percentage (%wt)
Leather	25
Polyurethane (PU)	17
Thermoplastic Rubber (TR)	16
Ethylene Vinyl Acetate (EVA)	14
Poly Vinyl Chloride (PVC)	8
Rubber	7
Other (Adhesives, Metals, etc.)	7
Textiles and Fabrics	6

Improvements in manufacturing technologies are attempting to control waste production from the source itself; however, complete elimination of waste is not possible and therefore proper end of life disposal is necessary. Since a shoe is composed of heterogeneous material, its end of life disposal is more complicated than that of commodities which are homogeneous. End of life techniques for shoes which have the least impact on the environment are:

(i) *Reuse*

When a shoe is disposed of mainly due to change in fashions or trends, the possibility of it being completely suitable for direct reuse is high. In some cases, however, minimal amendments can again make the disposed shoe ready for reuse. While this end of life disposal route seems to be the easiest, it requires a well-planned system that can ensure proper collection of such shoes and further distribute them.

(ii) *Recycling*

Footwear manufacturing consists of parts made with different materials which make recycling a complicated process. Recycling footwear can be widely classified as either a destructive method or a non-destructive method. The destructive recycling method involves shredding, which could alter the shoes into raw material for other useful products. Shredding is done without segregating different types of materials in a shoe. This mass of shredded material thus obtained is only suitable to be used as a filler material for levelling and surfacing of playgrounds, roads or footpaths. A famous example of such a destructive recycling programme in the footwear industry was Nike's 'Reuse-A-Shoe' programme (NIKE Reuse-A-Shoe, 2006).

Destructive recycling, however, limits the possible usage of end of life shoes and should be preferred only when other options are not feasible. Non-destructive recycling involves disassembling of footwear to acquire various components which did not undergo much wear and tear and can be easily reused. When the directly reusable components are separated out, the next step is to separate components on the basis of the recycling method and degree they require. Non-destructive recycling refers to disassembling of shoes to replace worn-out parts with new parts to repair that shoe or to make an entirely new product of the non-worn-out parts. However, the main restriction in popularising this method is the difficulty of dismantling the shoes. Thus, if this problem is rectified, this method will prove to be the most sustainable method for end of life footwear product management.

(iii) *Incineration and Landfilling*

Disposal of end of life footwear products in landfills is the least preferred route when its environmental impact is being taken into consideration. Strict regulations are now being implemented to avoid disposal of all types of waste in landfills unless absolutely necessary to circumvent the additional production of methane. Such waste is thus nowadays being used for energy recovery by incineration. Incineration of such materials produces large amounts of thermal energy which can be used to generate electricity. Apart from incineration, pyrolysis and gasification are increasingly being used for energy recovery. Especially when discarded footwear is made of leather, gasification can be used to generate heat and recover chromium as well.

2.3.2.2.1   Biodegradable Components of Footwear

Two well-known methods for managing end of life of biodegradable materials segregated from footwear are biological treatment and conventional methods. In the biological treatment method, both anaerobic and aerobic digestion are included. Treatment of biodegradable materials using aerobic digestion results in generation of water, methane and carbon dioxide along with some manure. This manure can act as a fertiliser. It is also reported that the processing time for composting can be decreased to a large extent if biodegradable materials are heated at 50–60°C. Some biodegradable materials also produce hazardous substances as by-products which can adversely affect the soil. Therefore, not all biodegradable materials extracted from shoes are suitable for composting and the materials should be meticulously screened before being sent for composting. For anaerobic digestion, a closed vessel lacking oxygen is used for degrading materials. In anaerobic digestion, biogas and digestate, a fertiliser, are produced as well as CO_2. If polylactide is one of the materials in the shoe being recycled, it will, however, behave as non-biodegradable in anaerobic conditions (Klauss & Bidlingmaier, 2004; Staikos et al., 2006). Figure 2.12 shows a flowchart for waste management options for shoes.

2.3.2.3   End of Life Disposal of Non-biodegradable Waste

2.3.2.3.1   End of Life Disposal of Automobiles

The number of people owning vehicles has increased at a rate higher than the increase of the global population (Hideto et al., 2014). Among all continents, Asia, Central America and South America have a clear trend for the rate of increase of vehicles. End of life vehicles (E.L.V.) have become a serious worldwide concern due to this increase. Recovery and recycling of the waste of E.L.V. are responsible for sustainable waste management (Seo & Kim, 2007). E.L.V.s contain more than 75% iron which can be recycled and reused easily through various market strategies, but oscillations in the market price of waste steel can alter the E.L.V. prices. Various efforts such as increasing the lifespan of vehicles, involving more ecofriendly materials, reducing gaseous emissions, etc. can be made to increase sustainability (BMW, 2000; Japan Automobile Manufacturers Association, 1999). Areas such as Europe and Japan have constraints in land availability for managing waste. Because of this, various guidelines regarding management of E.L.V.s are introduced which ensure an accurate recycling system. The aim of this strategy was to reduce the waste generation by up to 95% and to increase the rate of recycling (Endoh et al., 2006; EU, 2000. For the management of waste, E.L.V.s undergo a dismantling process to separate the useable parts. These parts are then sent to a shredding facility in compressed form. From the shredding process, Automobile Shredder Dust (A.S.D.) is produced which consists of ferrous and non-ferrous materials. Shredder Dust (S.D.) is also produced which contains dust and fluff (Seo & Kim, 2007). Almost all metallic materials are recycled after dismantling, whereas non-organic S.D.s are disposed of in landfills. Figure 2.13

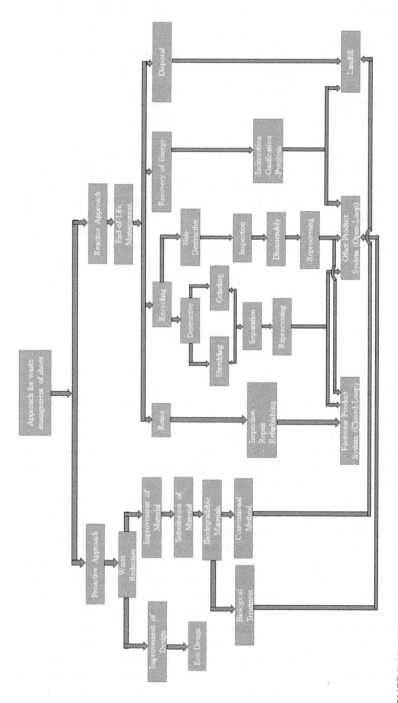

FIGURE 2.12 Waste management options for shoes (Staikos et al., 2006).

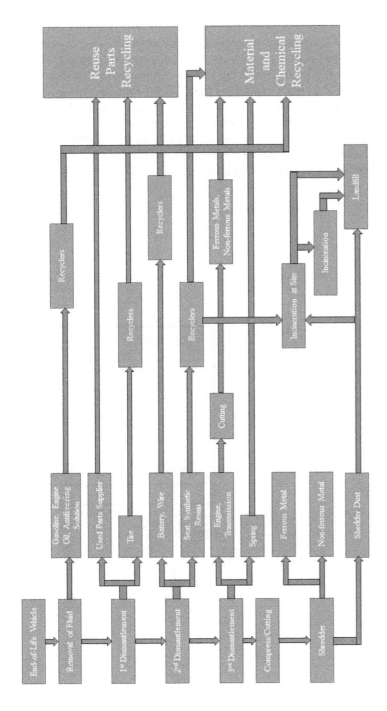

FIGURE 2.13 E.L.V. process (Seo & Kim, 2007).

shows the overall E.L.V. process, whereas Figure 2.14 shows the E.L.V. process in Japan.

(i) *End of Life (E.O.L.) Disposal of Automobiles in Australia*

There is no particular legislation for E.L.V. disposal in Australia. The management of E.L.V.s is carried out using economic processes (Soo et al., 2016). Harmful substances obtained from E.L.V. treatment are disposed of under specific arrangements provided by the Product Stewardship Act 2011 (The Parliament of the Commonwealth of Australia, 2011). Under this Act, voluntary communities seek authorisation from the government of Australia for their product steward arrangement, similarly to the Australian Initiative for Battery Recycling, Tyre Stewardship Australia and the Product

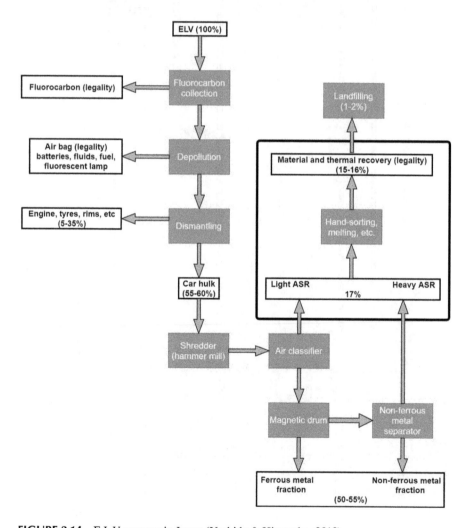

FIGURE 2.14 E.L.V. process in Japan (Yoshida & Hiratsuka, 2012).

Stewardship for Oil Program (ABRI, n.d.; Department of the Environment and Energy, Australia, n.d.). Hence, the recycling of batteries, tyres and fluids are carried out by these authorities. Also, the framework for product stewardship is implemented under the national waste policy (Department of the Environment, Australia, 2009). One disadvantage of the policy is competition between legal and illegal organisations. The illegal recycling organisations are not concerned about the environment and provide lower recycling prices when compared to legal organisations (McNamara, 2009). A flow chart for the E.L.V. process in Australia is given in Figure 2.15.

(ii) *End-of-Life Disposal of Automobiles in Belgium*

In Belgium, the management of E.L.V.s is based on the ELV Directive 200/53/EC passed in September 2000 (E.U. Directive, 2000). It includes various characteristics ranging from production of vehicles to different recycling stages on the basis of an extended producer responsibility policy (Hideto et al., 2014) and the subsidiarity principle (Smink, 2007). Under the subsidiarity principle, all the necessary guidelines of the directive are fulfilled on the basis of distinct approaches of Member States in their countries (Smink, 2007). The directive for E.L.V.s is executed at a regional level, and the monitoring process is carried out by a non-profit organisation named Febelauto. This organisation is responsible for the management of collection, recycling and treatment of E.L.V.s. Febelauto also informs and supports various groups like recycling operators, authorised treatment facilities and previous vehicle owners (Febelauto, n.d.). According to an important law in Belgium, various targets are given to vehicle recyclers for the recovery, reuse and recycling of E.L.V.s. The recovery and recycling efficiencies (η) can be stated as the ratio of the total mass produced from

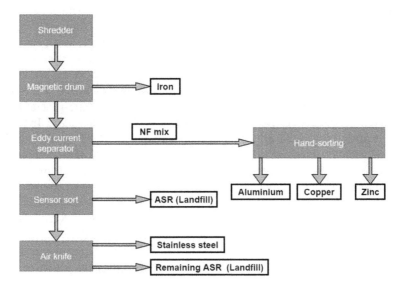

FIGURE 2.15   E.L.V. process in Australia (Soo et al., 2016).

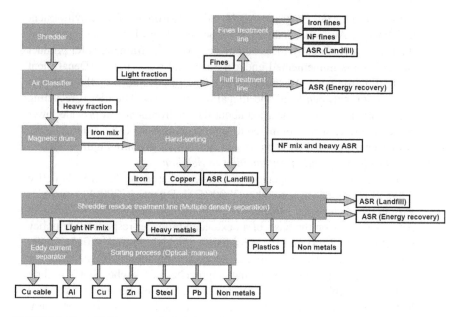

FIGURE 2.16   E.L.V. process in Belgium (Soo et al., 2016).

output material during the recycling process to the input. All the material losses obtained at the time of processing are included in the input during the efficiency calculation (Gerrard & Kandlikar, 2007). A flowchart showing the E.L.V. process in Belgium is given in Figure 2.16.

2.3.2.3.2   End of Life Disposal of Computers

According to a survey in the U.S., approximately 60 million computers are sold in the market and more than 12 million computers are disposed of annually, but only 10% of computers undergo the recycling process (Platt & Hyde, 1997; Ravi et al., 2005). The predicted amount of e-waste in three years from now includes one billion pounds of lead, four billion pounds of plastic, nearly two million pounds of cadmium, etc. (Silicon Valley Toxics Coalition, 2002). In a report issued by the U.S. National Safety Council, computers are ranked high in the list of solid waste-generating substances (Hamilton, 2001). These wastes are also known as solid e-waste. In the past, C.R.T. (Cathode Ray Tube) computer monitors were used, which are now replaced by L.E.D. monitors resulting in a reduction of the overall weight of the computer and composition of lead in e-waste (Puckett et al., 2005). The old C.R.T. monitors are recycled to a maximum extent, as recycled materials have a lower carbon footprint when compared to the raw materials used for the production of new products. The quality of the recycling process for E.O.L. computers depends on various factors, such as recovery of metal and recycling of plastics as well as C.R.T.

The process of recovery of materials begins after the elimination of harmful substances, working components and important parts. E-waste obtained from E.O.L. computers includes 6.3 wt% lead, 20.47 wt% iron, 6.93 wt% copper, 24.8 wt% glass, 23 wt% plastic, 0.02 wt% important metals, 14.17 wt% aluminium and 4.3 wt%

others (Kang & Schoenung, 2005). The recovery process can be carried out on the basis of reverse logistics for recycling of computer hardware. In this process, E.O.L. computers are received from the customers and a quality inspection is carried out to separate recyclable materials. In the absence of any recyclable material, proper disposal of material is carried out in such a way that it should not have any negative impact on sustainability. For recyclable materials, various recycling techniques are used and a new product/material is developed which can be further used in the manufacturing of computer parts (Ravi, 2012). A flowchart showing reverse logistics is given in Figure 2.17.

Apart from this, glass-to-glass and glass-to-lead technologies are particularly used for the recycling of C.R.T. computer monitors. In glass-to-lead recycling process, the recovery of copper and metallic lead from C.R.T. glass is carried out using the smelting process (Kang & Schoenung, 2005), whereas in the glass-to-glass recycling process, C.R.T.s collected from E.O.L. computers are sent to the recycling unit in which all the glass is broken into small pieces without separating the funnel glass and panel. The small pieces of glass are then transferred to manufacturers for the production of new C.R.T.s. The recycled glass pieces can be used as a raw material at lower prices compared to fresh material (Kang & Schoenung, 2005). In the case of recovery of metals, ferrous components are separated in a magnetic separator with the help of electric or permanent magnets. Among different magnetic separators, the overhead belt magnet separator is considered to be most appropriate to carry out this operation. It includes a belt containing ferrous materials and a collecting tank into

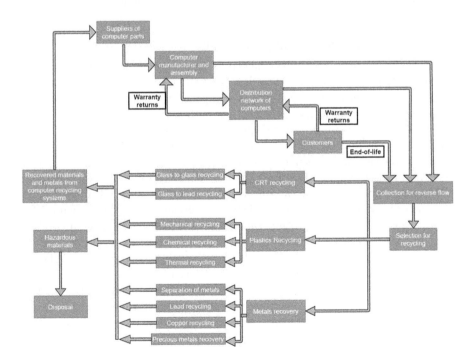

FIGURE 2.17 Layout for reverse logistics (Ravi 2012).

which non-ferrous materials are dropped due to gravity (Stessel, 2012). Non-ferrous metallic materials such as copper and aluminium collected in the tank can also be separated with the help eddy current separators and can be further used in manufacturing of various components. Therefore, various methods are used for the recycling and recovery of materials obtained from E.O.L. computers, which helps in maintaining sustainability to a large extent.

2.3.3   SUSTAINABLE WASTE MANAGEMENT – A NECESSITY FOR INDUSTRY 4.0

The primary point which has to be considered is that neoclassical business models cannot be supported sustainably. In order to bring sustainability, necessary changes have to be introduced, which should include environmental as well as social parameters (Stubbs & Cocklin, 2008). The major problem arises when a consumer demands a new product after the end of life for the old one instead of repairing or recycling the product (Guiltinan, 2009). A relationship between sustainable modernisation and the business model shows that four necessary elements of a business model are ignored. These models are supply chain, customer interface, financial model and value of proposition (De Man & Strandhagen, 2017). For some industries, maintaining sustainability and making profits is challenging, whereas some industries are in a strong financial position but lag behind when it comes to maintaining sustainability (Boons & Lüdeke-Freund, 2013). At present, every country is not only experiencing various issues related to sustainability, but also in innovations related to automation and digitalisation. This can be resolved by introducing a fourth industrial revolution, which is also known as Industry 4.0 (Stock & Seliger, 2016; Lele, 2019).

2.3.3.1   Important Elements of Industry 4.0

The most important element of this industry is regeneration or the recycling process for the waste products of an industry. For any type of production process, sustainability can be reached only when three general sustainable strategies are considered. These strategies are consistency, efficiency and sufficiency (De Man & Friege, 2016). Sufficiency can be described as reducing the input for production processes, and efficiency relates to economical use of resources during the process of extraction, transportation and transformation of a material, whereas consistency is defined as the recycling of waste output and reusing it in the production process which can therefore refer to sustainable waste management (De Man & Strandhagen, 2017). The process of sustainable production and sustainable waste management is given in Figure 2.18.

Other elements of Industry 4.0 are connectivity and interoperability (Rojko, 2017). This includes a continuous connection between manufacturing systems, devices and other components and machine-to-machine interaction (M.2.M) (Rojko, 2017). In this system, all the components are connected and interact via wireless connection; this is also known as I.o.T. This reduces the source of human error, because machines will be interacting to perform various tasks. Apart from machine interaction, human-machine interaction also plays a crucial role in the production system. This interaction comprises the collaborative work of human and robots to perform tedious and intricate tasks. Such processes are known as collaborative robotics (Guo et al., 2007; Veloso et al., 2012).

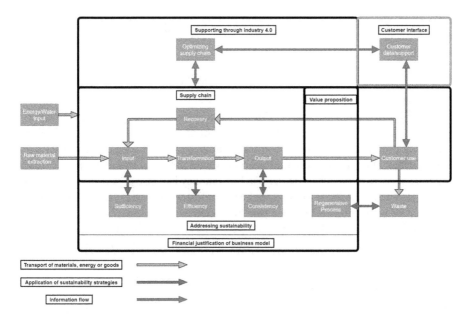

FIGURE 2.18 Connection between sustainable waste management and Industry 4.0 (De Man & Strandhagen, 2017).

2.3.3.2   Smart Industries

Smart industries are the factories with higher intelligence and flexibility, and they are more dynamic, which is a key feature for Industry 4.0. The process of production and waste management is observed with the help of automated systems and sensors. Because of the automated systems, the machines can optimise the production process and waste management according to the requirements (Dutta & Bose, 2015; Roblek et al., 2016). At present, there are two major topics of research in this field, which are intelligent production and the smart factory. In order to achieve the goal of converting older generation factories into smart factories, some necessary element must be considered. These elements are discussed below.

2.3.3.2.1   Efficient Management

When the size of older generation factories is compared to the size of smart factories, smart factories tend to have large infrastructures compared to older factories and much more complicated systems for production. Such huge systems must be managed efficiently. The management system for smart factories should be accurate and in order to achieve the same, an appropriate blueprint should be developed (Zhou et al., 2016).

2.3.3.2.2   Regular Training and Professional Development of the Staff

When it comes to smart factories, trends in technology tend to change very quickly and all the employees should be aware of upcoming technologies. For better understanding of technologies ensuring sustainable waste management, it becomes a responsibility of management to arrange training and workshops in this area. Such training is very useful when considering both technological and environmental factors. The workers will learn various necessary operations to make production more efficient and sustainable.

2.3.3.2.3   *Formation of a Trustworthy Industrial Communication System*

As discussed earlier, the infrastructure of a smart factory is complicated, and to make production efficient, having a high-quality communication system is a primary necessity. There may be different substations in an industry and production of any particular product needs a strong communication system between all the substations/departments. Having a bad communication system may lead to some major losses at the time of production. These losses can be of money because of wastage of material, or any physical injuries to workers. In heavy industries where cranes are used to lift and transfer heavy objects between two substations/departments, having an effective communication system during the entire process is imperative. Having trustworthy connectivity in an industry makes it safe which will lead to efficient production (Lucke et al., 2008).

2.3.3.2.4   *Safe, Secure and Sustainable Production*

In almost every industry, chemicals are used during the manufacturing of a product. For example, vehicle industries, medical industries, air conditioning and refrigeration industries, etc. use chemicals which can be harmful to humans as well as to the environment. In particular, if waste chemicals are not disposed of properly, then it may cause some major environmental issues. Such things are to be avoided to a maximum extent in smart factories. Apart from that, the workers should not feel any kind of threat to their lives. Some products are themselves dangerous if misused, and the security of such products is a very important aspect which should not be overlooked (Zhou et al., 2016).

2.3.3.2.5   *Controlled Input*

When compared to past industries, a lot of things have changed, including materials used for the production of a particular product, the technology used and other important parameters. All these factors can help to reduce the usage of resources to a great extent and provide efficient and sustainable production. Because of reduced amounts of resource usage, waste generation is minimal and therefore this results in sustainable waste management (Davis, 2017).

All parameters mentioned above are the key points which have to be considered for the creation of a smart industry.

2.3.3.3   Necessity of Industry 4.0

Nowadays, manufacturing processes have changed entirely due to the collaboration of digital technology and physical systems, and it is expected that this will further improve production in industries. After the implementation of Industry 4.0, mechanical processes are replaced by digital processes. Smart factories are part of Industry 4.0 and thus products manufactured in Industry 4.0 can be considered smart products. Considering present and future requirements, Industry 4.0 is a primary necessity for industries. These necessities are discussed in brief in the following subsections.

2.3.3.3.1   Possible Solution for Major Issues in Manufacturing

In today's world, the market is increasing rapidly which results in advancements in technologies. But there are certain issues, such as shorter time-cycles of products, complex production, market unpredictability and end of life disposal of products which should be resolved first. With the implementation of Industry 4.0, companies will not only become more active in business trends, but will also increase their flexibility. Industry 4.0 creates a path through which these issues can be solved easily. Observing the life-cycle of a product digitally, using production data to check its quality and getting feedback from customers can help in improving the life-cycle of a product. End of life disposal of a product can also be estimated with the help of acquired data. In Industry 4.0, the strength of the shop floor also increases because of smart machines and this will make production easier and more flexible. Sudden increase in demand is also an issue for which new production plans may be needed. This problem can be addressed easily with the help of digital production plans and re-planning can be implemented quickly.

2.3.3.3.2   Innovations in Production

Industry 4.0 not only helps in improving production efficiency, but also introduces new innovations in production. Industry 4.0 is already the latest revolution in industry and even a small innovation in this will lead to further improvements in production. Factories with smart machines can share their machines with other factories or even sell their smart machines which are not used for any purpose. On sharing the machines, other factories will also use the latest technologies which will affect their efficiency and production planning. Industry 4.0 is not just limited to production, but it can also be extended to servicing. Let us consider a vehicle of any particular company. For maintenance purposes, vehicles are sent to service stations instead of production plants. All the required digital equipment is available at the service stations with even more advanced technology when compared to manufacturing plants. These innovations have a major effect on the economy.

2.3.3.3.3   Effective Labour

With recent trends in technology, industries must focus on improving the skills of their employees. If the workforce is more productive, it will significantly increase the production rate (Blunck & Werthmann, 2017). New technologies proposed in Industry 4.0 will also improve the proficiency of workers (Mckinsey Digital, 2015). Let us take an example of two workers working in different factories, one of them in a smart factory and the other in a normal factory. Say that the initial production time of a product is 20 minutes, and after that it has to be transferred to another machine for processing. In case of the person working in the non-smart factory, the worker has to wait for 20 minutes before transferring the product to another machine, whereas the person working in a smart factory will not have to wait for the same time, as the product will be transferred automatically from one machine to another machine using robotic arms, or any other automatic source. Thus, from the above instance, it can be concluded that time as well as workload can be reduced in a smart factory which will lead to an increase in labour productivity.

2.3.3.3.4   Optimum Usage of Resources

Industry 4.0 provides flexibility to optimise the usage of resources. This will not only create possibilities for improving manufacturing processes, but will also help in sustainable waste management. Optimisation of resources can be achieved with the help of cyber-physical systems using real-time observations for a process. These technologies make the world automated, faster and more sustainable. Therefore, the reduction in material required for a particular process will lead to improvement in production by 3.5%, and the waste management will also be more sustainable (Mckinsey Digital, 2015).

2.3.3.3.5   Enabling Sustainable Prosperity

Old industries always had issues in dealing with certain factors such as economics, losses of jobs, shortages in resources due to overuse, and pollution. The concept of Industry 4.0 enables opportunities for improvement in all these factors. Usage of resources can be monitored digitally to avoid human error which will result in accurate usage of resources. Industry 4.0 brings new technology into the market. This increases the number of available job vacancies and thus improves the employment rate. Using optimum resources and introducing more job vacancies also improves the economy of a country. Moreover, the production will become innovative and smart. Since resource usage is optimum and production is more accurate, management of waste will therefore become easy and sustainable, resulting in eco-friendly production.

2.3.3.3.6   Improvement in Quality

Applications of Industry 4.0 emphasise product improvements as well as the quality of processes with the help of certain parameters such as advance process control and real-time error corrections to get stability in the manufacturing process (Mckinsey Digital, 2015). These parameters can help to increase cost savings which will further improve the quality of a product by 10–20%. This improvement in the quality of a product will decrease waste production and will therefore help in the sustainable management of waste. Consider an example of a coal-based steam power plant used for the generation of electricity. If high-quality coal is used for the generation of steam, then it will lead to an increase in steam production followed by a high electricity generation rate, therefore, less waste will be produced and sustainable waste management becomes easier. Similarly, Siemens decreased the rate of defective products to a minimum in order to maintain sustainability by using advanced technologies which were a part of Industry 4.0 (Blunck & Werthmann, 2017).

2.3.3.3.7   Coordination in Supply and Demand

Generation of waste can be prevented by introducing techniques which allow accurate co-ordination between customer and supplier. Demands of a customer should be perfectly met in terms of quality and quality. This can be achieved by using a simple technique known as crowd forecasting which is based on advance analytics. Such technologies can prevent mismatch between supply and demand (Mckinsey Digital, 2015). Therefore, it will increase accuracy and will bring synchronisation between supply and demand. Reduction in mismatch of supply and demand will minimise the rate of waste generation and production will become more sustainable.

2.3.3.4   Sustainable Manufacturing in Industry 4.0

According to the U.S. Department of Commerce, sustainable manufacturing can be defined as manufacturing of the types of products that will have minimum impact on the environment, conserve natural resources, save energy and are economical and safe for consumers, employees and communities (International Trade Administration. 2007). Moreover, sustainable manufacturing can also be defined as the 'designing of products and industrial systems in such a way that the natural resources and systems should not be degraded. It should ensure minimum effect of production on human beings, animals and environment' (Mihelcic et al., 2003). The manufacturing process is the most important operation in a chain of product development and supply. Because of this, it becomes necessary to bring sustainability to manufacturing, as it will yield sustainable products. This can be achieved by introducing 6R methodology into manufacturing. 6R stands for recycle, reuse, recover, remanufacture, redesign and reduce (Jawahir & Bradley, 2016). In the 6R methodology, the emphasis is on 'reduce' in the initial stages of manufacturing. 'Reduction' tends to involve using minimum raw material resources, as well as reduction of energy consumption at the time of manufacturing, whereas 'reuse' focuses on using a product or part of a product after its first life-cycle in such a way that it reduces the consumption of raw material. Reuse also involves the process of using old product materials together with fresh raw materials (U.S. Environmental Protection Agency, 2008) in order to decrease waste production. On the other hand, 'recycle' includes usage of products after their life-cycle to make some new or similar product. The products which undergo the recycling process are already waste products and cannot be used along with the raw material for manufacturing purposes. 'Recover' involves observing a product during its life-cycle and upgrading it at the time of its end of life-cycle (U.S. Environmental Protection Agency, 2008). In the 'recover' process, products can be disassembled, cleaned and modified to some extent and used again for a certain period of time. Some products are 'redesigned' on the basis of future requirements (Jayal et al., 2010). Figure 2.19 shows innovations in various manufacturing processes along with its influence on stakeholder value (Jayal et al., 2010) whereas Figure 2.20 shows the 6R closed loop system (Jaafar et al., 2007).

2.3.3.5   Advantages of Sustainable Manufacturing

Manufacturing processes can have a harmful effect on the environment, water, land and air. Sustainable manufacturing can not only minimise these hazardous effects but it can also improve the economy. Sustainable manufacturing has other positive impacts, such as a safe working atmosphere for employees and workers, less waste production and increased safety for people residing near to these industries. These are various advantages of sustainable manufacturing which will be discussed in the following subsections.

2.3.3.5.1   Reduced Energy Consumption

A survey conducted by the U.S. Energy Information Administration (E.I.A.) in 2006 gives an elaborated energy consumption report for manufacturing in terms of heating fuels, units of electricity consumed and usage of coal and coke for the period of

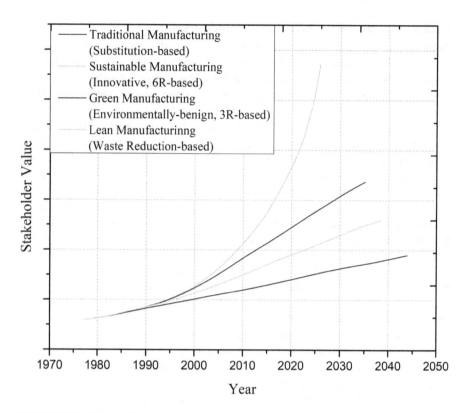

FIGURE 2.19 Innovations in manufacturing processes with respect to time (Jayal et al., 2010).

the previous five years (USEIA, 2006). It was reported that almost one-third of the energy produced is consumed in the U.S. Moreover, 61% of the produced energy is consumed by the refining industry, the paper industry, the steel industry, the food industry and the chemical industry, whereas 25% of energy is consumed during shipments of the product (USEIA, 2010). But it is necessary to note that energy consumption is not due to the manufacturing process. Instead more energy is consumed during the usage of the product. This implies that in order to reduce energy consumption, the product has to be more efficient and the manufacturing process should be sustainable. Consider the example of an industry which manufactures air conditioners. Making the manufacturing process sustainable will not only minimise the negative impact on the environment, but the machining process will also consume less energy. Therefore, efficient air-conditioning will reduce H.V.A.C. (heating, ventilation and air conditioning) consumption which results in a decrease in the greenhouse effect and an increase in sustainability (Haapala et al., 2013).

2.3.3.5.2 Reduced Consumption and Wastage of Water

There are many manufacturing processes that consume huge amounts of water. Producing even a single newspaper consumes approximately 250 gal of water, whereas production of a car can consume up to 10,000 gal of water (Masters & Ela, 2008).

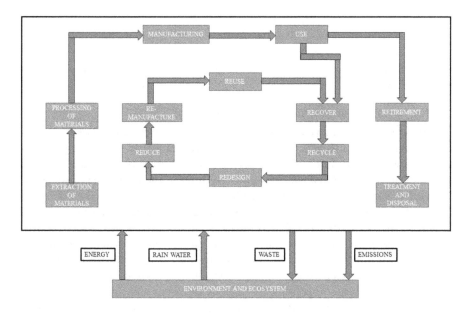

FIGURE 2.20 Closed loop system involving 6R methodology (Jaafar et al., 2007).

This consumption includes various parameters such as cleaning, cooling and quenching. In the food industry, a large amount of water is wasted during manufacturing of a product. Also, many manufacturing processes are responsible for water pollution. The most hazardous substances produced in a manufacturing process which causes water pollution are heavy metals and VOCs (Haapala et al., 2013). Apart from these substances, grease, oil and biochemical products used during manufacturing also harmful. These substances dissolve in water and cause water pollution. Therefore, sustainable manufacturing will eliminate the usage of unnecessary water and water pollution. Water purification systems can also be implemented for waste-water in such a way that it can be used again in manufacturing. Machines with the latest technology are being used for manufacturing purposes, and consume the minimum amount of water and thus reduce water consumption. Non-toxic substances are also used which decreases water pollution and thus results in easier water purification.

2.3.3.5.3 Reduction in Solid Waste

During the manufacturing process, solid waste products are an unavoidable leftover, and their size can range from small flakes to large blocks. Increases in policies and goods prices resulted in developments in zero-waste manufacturing. Industries such as General Motors, Proctor and Gamble and Honda are operating zero-waste manufacturing amenities. Production in zero-waste facilities is considered to be sustainable because it decreases the negative impact on the environment due to reductions in generated waste and consumption of resources (Moreira et al., 2010). In most cases, reducing the waste generated during production is the initial emphasis, whereas recycling of the inevitable waste is a secondary focus. If there is no provision for minimising the waste or recycling, then production of energy is considered to be another option for sustainable manufacturing. General Motors has given an example of sustainable waste management by creating sound absorbers by recycling cardboard material used for shipping of products (Haapala et al., 2013). In this way, solid waste generated at the time of manufacturing is reduced and therefore it improves sustainable waste management (U.S. Environmental Protection Agency, 2003).

2.3.3.5.4   Reduction in Gaseous Emissions

Harmful gases emitted into the atmosphere can have adverse effects on the environment and on human beings (Rivera et al., 2007). Harmful particulate substances present in the atmosphere cause various diseases such as silicosis, asthma, lung cancer, urinary tract infections, emphysema and diseases of the larynx. These emissions also include greenhouse gases which are emitted due to the combustion of coal, usage of energy and etching of semiconductors. It also consists of gases emitted from ozone-depleting chemicals such as propellants, foam insulation and refrigerants which are very hazardous for the ozone layer. Chemicals like cleaning solvents and paint fumes are also responsible for airborne emissions. Gaseous emissions are also caused by chemicals which form smog, such as volatile organic compounds (V.O.C.s) and nitrogen oxides. All these chemicals are responsible for gaseous emissions and can degrade environmental performance. Today there are many eco-friendly refrigerants which do not emit harmful gases, and high-quality coal and the latest technology machines used in industries decrease the gaseous emission to a large extent. This brings sustainability in manufacturing and therefore decreases the emission of harmful gases into the atmosphere.

2.3.4   CONCLUSION

In this chapter, an in-depth detailed analysis is carried out of end of life disposal of various biodegradable commodities such as textiles, dairy products and biodegradable plastic. Also, a detailed study is performed on end of life disposal of non-biodegradable products like computers and automobiles. In end of life disposal of computers, a brief explanation is provided for different recycling and recovery methods. The end of life disposal of automobiles is explained with the help of all necessary flowcharts for various continents. Apart from end of life disposal of various products, management of waste in Industry 4.0 is explained thoroughly. This includes an overview of the necessity for Industry 4.0 followed by a short introduction of important elements of Industry 4.0 and smart factories. In reference to smart factories, various parameters are discussed which are required for the conversion of an old factory into a smart factory. At the end, some important points are highlighted that are required to make manufacturing more sustainable, along with some advantages of Industry 4.0.

REFERENCES

Abbot, S., and A. Wilford. "The Footwear Industry and the Environment, Modern Shoemaking."
      No. 56. Kettering: SATRA. 1999.
ABRI. (n.d.) Australian Battery Recycling Initiative. http://www.batteryrecycling.org.au/
      home. (Accessed August 16, 2016).

Aly, H. M., and A. A. M. Daifullah. 1998. "Potential Use of Bagasse Pith for the Treatment of Wastewater Containing Metals." *Adsorption Science and Technology* 16(1): 33–38. doi:10.1177/026361749801600105.

Aminabhavi, T. M, R. H. Balundgi, and P. E. Cassidy. 2017. "A Review on Biodegradable Plastics." *Polymer-Plastics Technology and Engineering* 29(3): 235–262. doi:10.1080/03602559008049843.

Bhardwaj, Vertica, and Ann Fairhurst. 2010. "The International Review of Retail, Distribution and Consumer Research Fast Fashion : Response to Changes in the Fashion Industry." *The International Review of Retail, Distribution and Consumer Research* 20: 37–41. doi:10.1080/09593960903498300.

Birtwistle, G, and C. M. Moore. 2007. "Fashion Clothing – Where Does It All End Up ?" *International Journal of Retail & Distribution Management* 35(3): 210–216. doi:10.1108/09590550710735068.

Bloom, David E., David Canning, and Pia N. Malaney. 2000. "Population Dynamics and Economic Growth in Asia." *Population & Development Review* 26(May): 257–290.

Bloom, David E., and Richard B. Freeman. 1988. "Economic Development and the Timing and Components of Population Growth." *Journal of Policy Modeling* 10(1): 57–81. doi:10.1016/0161-8938(88)90035-X.

Blunck, Erskin, and Hedwig Werthmann. 2017. "Industry 4.0 - An Opportunity to Realize Sustainable Manufacturing and Its Potential for a Circular Economy." *Microeconomics* 3(2): 645–666.

BMW. 2000. BMW Environmental Report, 1997/98–2000. Muenchen, Germany.

Boons, Frank, and Florian Lüdeke-Freund. 2013. "Business Models for Sustainable Innovation: State-of-the-Art and Steps towards a Research Agenda." *Journal of Cleaner Production* 45: 9–19. doi:10.1016/j.jclepro.2012.07.007.

British Footwear Association. 2005. http://Www.Britfoot.Com (Accessed December 15, 2005)."

Cantrell, Keri B., Thomas Ducey, Kyoung S. Ro, and Patrick G. Hunt. 2008. "Bioresource Technology Livestock Waste-to-Bioenergy Generation Opportunities." *Bioresource Technology* 99: 7941–7953. doi:10.1016/j.biortech.2008.02.061.

Cassells, Sue, John Holland, and Anton Meister. 2005. "End-of-Life Vehicle Disposal: Policy Proposals to Resolve an Environmental Issue in New Zealand." *Journal of Environmental Policy and Planning* 7(2): 107–124. doi:10.1080/15239080500338499.

Chan, Joseph W. K., and Thomas K. L. Tong. 2007. "Multi-Criteria Material Selections and End-of-Life Product Strategy: Grey Relational Analysis Approach." *Materials and Design* 28(5): 1539–1546. doi:10.1016/j.matdes.2006.02.016.

Council Directive 1999/31/EC of 26 April 1999 on the Landfill of Waste.

Davis, Jim. 2017. "Smart Manufacturing." *Encyclopedia of Sustainable Technologies* 7543(July): 417–427. doi:10.1016/B978-0-12-409548-9.10212-X.

Davis, J. J., R. W. Perkins, R. F. Palmer, W. C. Hanson, and J. E. Cline. 1958. "Radioactive Materials in Aquatic and Terrestrial Organisms Exposed to Reactor Effluent Water."

Department of the Environment and Energy, Australia. (n.d.). "Product Stewardship for Oil (PSO) Program." https://www.environment. gov.au/protection/used-oil-recycling/product-stewardship-oilprogram (Accessed August 16, 2016).

Department of the Environment, Australia. 2009. "National Waste Policy: Less Waste, More Resources. Natl Waste Policy Fact Sheet 2009." http://www.environment.gov.au/system/files/pages/94aa70c5-668144c6-8d83-77606d1d6afe/files/fs-national-waste-policy.pdf.

Dutta, Debprotim, and Indranil Bose. 2015. "Managing a Big Data Project: The Case of Ramco Cements Limited." *International Journal of Production Economics* 165: 293–306. doi:10.1016/j.ijpe.2014.12.032.

E.U. Directive. 2000. 53/EC of the European Parliament and of the Council of 18 September 2000 on End-of-Life Vehicles. *Off J Eur Union Ser* 2000:34–42.

Ekvall, Tomas. 2001. "Allocation in ISO 14041 — a Critical Review." *Journal of Cleaner Production* 9: 197–208.

Endoh, Shigehisa, Kenzo Takahashi, Jae Ryeong Lee, and Hitoshi Ohya. 2006. "Mechanical Treatment of Automobile Shredder Residue for Its Application as a Fuel." *Journal of Material Cycles and Waste Management* 8(1): 88–94. doi:10.1007/s10163-005-0140-7.

EU. 2000. 2000/53/EC. Directive of European Parliament and of the Council.

Febelauto. (n.d.) http://www.febelauto.be/en/ (Accessed August 17, 2016).

Garcia-Garcia, Guillermo, Elliot Woolley, and Shahin Rahimifard. 2016. "A Framework for a More Efficient Approach to Food Waste Management." *ETP International Journal of Food Engineering*. doi:10.18178/ijfe.1.1.65-72.

Gerrard, Jason, and Milind Kandlikar. 2007. "Is European End-of-Life Vehicle Legislation Living up to Expectations? Assessing the Impact of the ELV Directive on 'green' Innovation and Vehicle Recovery." *Journal of Cleaner Production* 15(1): 17–27. doi:10.1016/j.jclepro.2005.06.004.

Ghimire, Anish, Luigi Frunzo, Francesco Pirozzi, Eric Trably, Renaud Escudie, Piet N. L. Lens, and Giovanni Esposito. 2015. "A Review on Dark Fermentative Biohydrogen Production from Organic Biomass : Process Parameters and Use of by-Products." *Applied Energy* 144: 73–95. doi:10.1016/j.apenergy.2015.01.045.

Greyson, James. 2007. "An Economic Instrument for Zero Waste, Economic Growth and Sustainability." *Journal of Cleaner Production* 15(13–14): 1382–1390. doi:10.1016/j.jclepro.2006.07.019.

Gross, Richard A., and Bhanu Kalra. 2002. "Biodegradable Polymers for the Environment." *Science* 297(4): 803–808.

Guiltinan, Joseph. 2009. "Creative Destruction and Destructive Creations: Environmental Ethics and Planned Obsolescence." *Journal of Business Ethics* 89(SUPPL. 1): 19–28. doi:10.1007/s10551-008-9907-9.

Guo, Yi, Lynne E. Parker, and Raj Madhavan. 2007. "Collaborative Robots for Infrastructure Security Applications." *Studies in Computational Intelligence* 50: 185–200. doi:10.1007/978-3-540-49720-2_9.

Haapala, Karl R., Fu Zhao, Jaime Camelio, John W. Sutherland, Steven J. Skerlos, David A. Dornfeld, I. S. Jawahir, Andres F. Clarens, and Jeremy L. Rickli. 2013. "A Review of Engineering Research in Sustainable Manufacturing." *Journal of Manufacturing Science and Engineering* 135(4): 041013. doi:10.1115/1.4024040.

Hamilton, A. 2001. "How Do You Junk Your Computer?" *Time* 157(6): 70–71.

Handl, Günther. 2012. "Declaration of the United Nations Conference on the Human Environment (Stockholm Declaration), 1972 and the Rio Declaration on Environment and Development, 1992." *United Nations Audiovisual Library of International Law*, 1–11.

Harvey, A. J. 1982. "Footwear Materials and Process Technology. New Zealand Leather and Shoe Research Association."

Hideto, Shin-ichi Sakai, Yoshida Jiro, Hiratsuka Carlo, and Vandecasteele Regina. 2014. "An International Comparative Study of End-of-Life Vehicle (ELV) Recycling Systems." *Journal of Material Cycles and Waste Management*, 1–20. doi:10.1007/s10163-013-0173-2.

Hsiang, Solomon M., and Marshall Burke. 2014. "Climate, Conflict, and Social Stability: What Does the Evidence Say?" *Climatic Change* 123(1): 39–55. doi:10.1007/s10584-013-0868-3.

International Trade Administration. 2007. "How Does Commerce Define Sustainable Manufacturing? U.S. Department of Commerce." http://%0Awww.trade.gov/competitiveness/sustainablemanufacturing/how:doc_defines_%0ASM.asp.

Itorial, E. D. 2017. "To Launch The." 358(6365). doi:10.1126/science.aao6749.

Jaafar, I. H., A. Venkatachalam, K. Joshi, A. C. Ungureanu, N. De Silva, O. W. Dillon Jr, and K. E. Rouch, and I. S. Jawahir. 2007. "Product Design for Sustainability: A

New Assessment Methodology and Case Studies." In Kutz, M, (Ed.) *Handbook of Environmentally Conscious Mechanical Design*. New Jersey: John Wiley & Sons.

Japan Automobile Manufacturers Association. 1999. "Japan Automobile Manufacturers Association Strategy of Recycling ELVs." JAMA Report, Vol 49, JAMA.

Jawahir, I. S., and Ryan Bradley. 2016. "Technological Elements of Circular Economy and the Principles of 6R-Based Closed-Loop Material Flow in Sustainable Manufacturing." *Procedia CIRP* 40: 103–108. doi:10.1016/j.procir.2016.01.067.

Jayal, A. D., F. Badurdeen, O. W. Dillon, and I. S. Jawahir. 2010. "Sustainable Manufacturing: Modeling and Optimization Challenges at the Product, Process and System Levels." *CIRP Journal of Manufacturing Science and Technology* 2(3): 144–152. doi:10.1016/j. cirpj.2010.03.006.

Kang, Hai Yong, and Julie M. Schoenung. 2005. "Electronic Waste Recycling: A Review of U.S. Infrastructure and Technology Options." *Resources, Conservation and Recycling* 45(4): 368–400. doi:10.1016/j.resconrec.2005.06.001.

Klauss, M., and W. Bidlingmaier. 2004. "Biodegradable Polymer Packaging: Practical Experiences of the Model Project Kassel." In *Proceedings of the 1st UK Conference and Exhibition on Biodegradable and Residual Waste Management*, Harrogate, Leeds, February (pp. 18–19).

Koligkioni, Athina, Keshav Parajuly, Birgitte Liholt Sørensen, and Ciprian Cimpan. 2018. "Environmental Assessment of End-of-Life Textiles in Denmark." *Procedia CIRP* 69(May): 962–967. doi:10.1016/j.procir.2017.11.090.

Kónya, József, and Noémi M. Nagy. 2018. "Environmental Radioactivity." *Nuclear and Radiochemistry*, 399–419. doi:10.1016/b978-0-12-813643-0.00013-5.

Lele, Ajey. 2019. "Industry 4.0." *Smart Innovation, Systems and Technologies* 132: 205–215. doi:10.1007/978-981-13-3384-2_13.

Lenzing. 2017. "The Global Fiber Market in 2016.".

Lom, Michal, Ondrej Pribyl, and Miroslav Svitek. 2016. "Industry 4.0 as a Part of Smart Cities BT - 2016 Smart Cities Symposium Prague, SCSP 2016, May 26, 2016 - May 27, 2016." *Smart Cities Symposium Prague*, 2–7. doi:10.1109/SCSP.2016.7501015.

Lucas, Robert E. 1989. "On the mechanics of economic development." NBER Working Paper R1176.

Lucke, Dominik, Carmen Constantinescu, and Engelbert Westkämper. 2008. "Smart Factory - A Step towards the Next Generation of Manufacturing." *Manufacturing Systems and Technologies for the New Frontier* Sfb 627: 115–118. doi:10.1007/978-1-84800-267-8_23.

Lutz, Wolfgang, Warren Sanderson, and Sergei Scherbov. 1997. "Doubling of World Population Unlikely." *Nature* 387(6635): 803–805. doi:10.1038/42935.

Lyberatos, I., G. Ntaikou, and G. Antonopoulou. 2010. "Biohydrogen Production from Biomass and Wastes via Dark Fermentation : A Review." *Waste and Biomass Valorization* 21–39. doi:10.1007/s12649-009-9001-2.

Man, Johannes Cornelis De, and Jan Ola Strandhagen. 2017. "An Industry 4.0 Research Agenda for Sustainable Business Models." *Procedia CIRP* 63: 721–726. doi:10.1016/j. procir.2017.03.315.

Man, Reinier De, and Henning Friege. 2016. "Circular Economy: European Policy on Shaky Ground." *Waste Management and Research* 34(2): 93–95. doi:10.1177/0734242X15626015.

Masters, Gilbert M., and Wendell P. Ela. 2008. *Introduction to Environmental Engineering and Science. No. 60457*. Englewood Cliffs, NJ: Prentice Hall.

Mckinsey Digital. 2015. "Industry 4.0 How to Navigate Digitization of the Manufacturing Sector." *Mckinsey Digital*. doi:10.1080/18811248.1966.9732270.

McNamara, Nova Elizabeth. 2009. "Vehicle Recycling and Sustainability." October: 80. http://www.toyota-global.com/sustainability/report/vehicle_recycling/pdf/vr_all.pdf.

Michal Perle, Shlomo Kimchie, and Gedaliah Shelef. 1995. "Some Biochemical Aspects of the Anaerobic Degradation of Dairy Wastewater Michal."

Mihelcic, J. R., J. C. Crittenden, M. J. Small, D. R. Shonnard, D. R. Hokanson, Q. Zhang, H. Chen, S. A. Sorby, V. U. James, J. W. Sutherland, and J. L. Schnoor. 2003. "Sustainability Science and Engineering: The Emergence of a New Metadiscipline." *Environmental Science & Technology*. doi:10.1021/es034605h.

Moreira, F., A. C. Alves, and R. M. Sousa. 2010. "Towards Eco-Efficient Lean Production Systems, IFIP Advances in Information and Communication Technology." doi:10.1007/978-3-642-14341-0_12.

Mudgal, S, K. Muehmel, E. Hoa, and M. Grémont. 2012. "Options to Improve the Biodegradability Requirements in the Packaging Directive."

NIKE Reuse-A-Shoe. 2006. http://Www.Nike.Com (Accessed January 15 2006).

PlasticsEurope. 2015. "Plastics – the Facts 2015, An Analysis of European Plastics Production, Demand and Waste Data. PlasticsEurope."

Platt. B, and J. Hyde. 1997. "Plug into Electronics Reuse. Institute for Local Self-Reliance." January.

Puckett, J., S. Westervelt, R. Gutierrez, and Y. Takamiya. 2005. "The Digital Dump. Exporting Re-Use and Abuse to Africa." Report from the Basel Action Network.

Ravi, V. 2012. "Evaluating Overall Quality of Recycling of E-Waste from End-of-Life Computers." *Journal of Cleaner Production* 20(1): 145–151. doi:10.1016/j.jclepro.2011.08.003.

Ravi, V., Ravi Shankar, and M. K. Tiwari. 2005. "Analyzing Alternatives in Reverse Logistics for End-of-Life Computers: ANP and Balanced Scorecard Approach." *Computers and Industrial Engineering* 48(2): 327–356. doi:10.1016/j.cie.2005.01.017.

Reimer, B., M. S. Sodhi, and W. A. Knight. 2002. "Optimizing Electronics End-of-Life Disposal Costs." In *Proceedings of the 2000 IEEE International Symposium on Electronics and the Environment (Cat. No. 00CH37082)* (pp. 342–347). doi:10.1109/isee.2000.857672.

Rivera, J. L., D. J. Michalek, and J. W. Sutherland. 2007. "Air Quality in Manufacturing." *Environmentally Conscious Manufacturing*. New Jersey: Wiley, pp. 145–178.

Roblek, Vasja, Maja Meško, and Alojz Krapež. 2016. "A Complex View of Industry 4.0." *SAGE Open* 6(2). doi:10.1177/2158244016653987.

Rojko, Andreja. 2017. "Industry 4.0 Concept: Background and Overview." *International Journal of Interactive Mobile Technologies* (iJIM) 11(5): 14.

Roos, Sandra, Stefan Posner, Greg M. Peters, and Christina Jo. 2015. "Is Unbleached Cotton Better Than Bleached ? Exploring the Limits of Life-Cycle Assessment in the Textile Sector." *Clothing and Textiles Research Journal* 33(4): 231–247. doi:10.1177/0887302X15576404.

Rujnić-Sokele, Maja, and Ana Pilipović. 2017. "Challenges and Opportunities of Biodegradable Plastics: A Mini Review." *Waste Management and Research* 35(2): 132–140. doi:10.1177/0734242X16683272.

Sandin, Gustav, and Greg M. Peters. 2018. "Environmental Impact of Textile Reuse and Recycling e A Review." *Journal of Cleaner Production* 184: 353–365. doi:10.1016/j.jclepro.2018.02.266.

Sandin, Gustav, Greg M. Peters, and Magdalena Svanström. 2015. "Using the Planetary Boundaries Framework for Setting Impact-Reduction Targets in LCA Contexts." *The International Journal of Life Cycle Assessment*. doi:10.1007/s11367-015-0984-6.

Schmidt, A., D. Watson, S. Roos, C. Askham, and P. B. Poulsen. 2016. "Gaining Benefits from Discarded Textiles: LCA of Different Treatment Pathways."

Seo, Hyun-tae Joung Sung-jin Cho Yong-chil, and Woo-hyun Kim. 2007. "Status of Recycling End-of-Life Vehicles and Efforts to Reduce Automobile Shredder Residues in Korea." *Journal of Material Cycles and Waste Management*, 159–166. doi:10.1007/s10163-007-0181-1.

Silicon Valley Toxics Coalition. 2002. "Take It Back! Make It Clean! Make It Green! Computer Takeback Campaign." http:// Www.Svtc.Org/Cleancc/Pubs/2002report.Htm."

Smink, Carla K. 2007. "Vehicle Recycling Regulations: Lessons from Denmark." *Journal of Cleaner Production* 15(11–12): 1135–1146. doi:10.1016/j.jclepro.2006.05.028.

Song, J. H., R. J. Murphy, R. Narayan, G. B. H. Davies, J. H. Song, R. J. Murphy, R. Narayan, and G. B. H. Davies. 2009. "Biodegradable and Compostable Alternatives to Conventional Plastics Biodegradable and Compostable Alternatives to Conventional Plastics." *Philosophical Transactions of the Royal Society B: Biological Sciences.* doi:10.1098/rstb.2008.0289.

Soo, Vi Kie, Paul Compston, and Matthew Doolan. 2016. "Is the Australian Automotive Recycling Industry Heading towards a Global Circular Economy? - A Case Study on Vehicle Doors." *Procedia CIRP* 48: 10–15. doi:10.1016/j.procir.2016.03.099.

Staikos, T., and S. Rahimifard. 2007. "An End-of-Life Decision Support Tool for Product Recovery Considerations in the Footwear Industry." *International Journal of Computer Integrated Manufacturing* 20(6): 602–615. doi:10.1080/09511920701416549.

Staikos, Theodoros, Richard Heath, Barry Haworth, and Shahin Rahimifard. 2006. "End-of-Life Management of Shoes and the Role of Biodegradable Materials." *End-of-Life Management of Shoes and the Role of Biodegradable Materials Theodoros*, 497–502.

Stavropoulos, K. P., A. Kopsahelis, C. Zafiri, and M. Kornaros. 2016. "Effect of PH on Continuous Biohydrogen Production from End-of-Life Dairy Products (EoL-DPs) via Dark Fermentation." *Waste and Biomass Valorization* 7(4): 753–764. doi:10.1007/s12649-016-9548-7.

Steffen, Will, Katherine Richardson, Johan Rockström, Sarah E. Cornell, Ingo Fetzer, Elena M. Bennett, Reinette Biggs, S. R. Carpenter, W. D. Vries, C. A. D. Wit, and C. Folke, C. 2015. "Planetary Boundaries : Guiding Changing Planet." *Science* 1259855. doi:10.1126/science.1259855.

Stessel, R. I. 2012. *Recycling and Resource Recovery Engineering: Principles of Waste Processing.* Berlin, Germany: Springer Science & Business Media.

Stock, T., and G. Seliger. 2016. "Opportunities of Sustainable Manufacturing in Industry 4.0." *Procedia CIRP* 40(Icc): 536–541. doi:10.1016/j.procir.2016.01.129.

Stubbs, Wendy, and Chris Cocklin. 2008. "Conceptualizing a 'Sustainability Business Model.'" *Organization and Environment* 21(2): 103–127. doi:10.1177/1086026608318042.

Taylor, Publisher, F. Gironi, and V. Piemonte. 2011. "Energy Sources, Part A : Recovery, Utilization, and Environmental Effects Bioplastics and Petroleum-Based Plastics : Strengths and Weaknesses Bioplastics and Petroleum-Based Plastics : Strengths and Weaknesses." *Energy Sources, Part A: Recovery, Utilization, and Environmental Effects* 33: 37–41. doi:10.1080/15567030903436830.

The Parliament of the Commonwealth of Australia. Product Stewardship Act 2011.

Thumerelle, P. J. 2001. "National Research Council: Beyond Six Billion, Forecasting the World's Population."

Toffel, Michael W. 2003. "The growing strategic importance of end-of-life product management." *California Management Review* 45(3): 102–129.

Treloar, Andrew. 2004. "Assessing Environmental Impacts of Manufacturing : Technologies for Informing Design." November 2018.

U.S. Environmental Protection Agency. 2003. Lean Manufacturing and The Environment: Research on Advanced Manufacturing Systems and the Environment and Recommendations for Leveraging Better Environmental Performance. *US EPA.*

U.S. Environmental Protection Agency. 2008. "Municipal Solid Waste (MSW) – Reduce, Reuse, and Recycle." http://www.epa.gov/msw/reduce.%0Ahtm.

UNIDO. 2000. "Wastes Generated in the Leather Footwear Industry, 14th Session of the Leather and Leather Products Industry Panel, Czech Republic."

USEIA. 2006. "Energy Use in Manufacturing 1998 to 2002." U.S. Department of Energy, U.S. Energy Information Administration. http://www.eia.doe. gov/emeu/mecs/special_top ics/energy_use_manufacturing/energyuse98_02/ 98energyuseO2.html.

USEIA. 2010. "Annual Energy Outlook 2011 Early Release Overview." U.S. Energy Information Administration. http://www.eia.gov/forecasts/aeo/early_ consumption.cfm.

Veloso, Manuela, Joydeep Biswas, Brian Coltin, Stephanie Rosenthal, Tom Kollar, Cetin Mericli, Mehdi Samadi, Susana Brandao, and Rodrigo Ventura. 2012. "CoBots: Collaborative Robots Servicing Multi-Floor Buildings." *IEEE International Conference on Intelligent Robots and Systems*, 5446–5447. doi:10.1109/IROS.2012.6386300.

Wang, Youjiang. 2010. "Fiber and Textile Waste Utilization." *Waste and Biomass Valorization*, 135–143. doi:10.1007/s12649-009-9005-y.

World Footwear. 2005. *The Future of Polyurethane Soling, World Footwear*. Cambridge, MA: Shoe Trades, pp.18–20.

Yee, Loi Wai, Siti Hasnah Hassan, and T. Ramayah. 2016. "Sustainability and Philanthropic Awareness in Clothing Disposal Behavior Among Young Malaysian Consumers." *Sage Open*. doi:10.1177/2158244015625327.

Yoshida, H., J. Hiratsuka. 2012. "Overview and Current Status of ELV Recycling in Japan." In *International Workshop on 3R Strategy and ELV Recycling*, September (pp. 19–21).

Zamani, Bahareh, Gregory Peters, and Tomas Rydberg. 2014. "A Carbon Footprint of Textile Recycling A Case Study in Sweden" *Journal of Industrial Ecology* 1–12. doi:10.1111/ jiec.12208.

Zhou, Keliang, Taigang Liu, and Lifeng Zhou. 2016. "Industry 4.0: Towards Future Industrial Opportunities and Challenges." In *2015 12th International Conference on Fuzzy Systems and Knowledge Discovery, FSKD 2015* pp. (2147–2152). doi:10.1109/ FSKD.2015.7382284.

3 Innovation for Smart Factories

3.1 ROLE OF INDUSTRIAL INTERNET OF THINGS (I.I.O.T.) MANUFACTURING

Lokesh Singh, Someh Kumar Dewangan, Ashish Das, and K. Jayakrishna

3.1.1 INTRODUCTION TO THE ROLE OF THE INDUSTRIAL INTERNET OF THINGS (I.I.O.T.) MANUFACTURING

The industrial revolution introduced a period in which individuals and machines started cooperating on speed generation times, improve quality and generally driving process efficiencies (Boyes et al., 2018).

Manufacturers will make capital investments in innovation, which will, thus, lead to long-term reductions in operational costs. Networked machines, sensors, etc. do not commit errors or require breaks (Jeschke et al., 2017) or preparation, and in this way they offer a solid and financially savvy approach to improve efficiency (Dalenogare et al., 2018).

The qualification clearly is fairly artificial, and in all dimensions, there are overlaps (Geisbauer et al., 2016). The quickest developing classifications of the internet of things (I.o.T.) use cases, for example, are cross-industry. In addition, albeit with a few advances, engineering systems and applications over all I.o.T. show a contrast (edge computing and fog processing are ordinary in Industrial I.o.T., there are distinctive sorts of system and availability apparatuses, I.I.o.T. passages fill different needs, Industrial I.o.T. stages bolster different use cases to I.o.T. stages, by and large, computerised twins are basically about industrial markets, the use cases for increased truth condition of industrial market are not equivalent, etc.). Between Industrial I.o.T. and Consumer I.o.T. a normal vast I.I.o.T. undertaking will use a few types of network and arrangements, some of which are utilised in customer I.o.T. too.

3.1.1.1 Evolution of I.I.o.T. in Industry

In order to move up in the I.I.o.T. development and possibility/opportunity reality, industrial organisations clearly need to begin somewhere (Hermann et al., 2015). It is usual that in the beginning stages, networking in the I.I.o.T. space is concentrated on a limited arrangement of objectives and advantages. However, it is essential to have a guide or plan for the more drawn out term. It is anything but a coincidence that all holistic challenges we find in the advancement of I.I.o.T. are actually equivalent to the ones we find in the computerised changes of assembly, the fundamental I.I.o.T. market (Dorsemaine et al., 2016).

In this Industry 4.0 or 'Industrial Internet' setting, where we essentially discover the I.I.o.T. as a major aspect of an incorporated methodology where it becomes the overwhelming focus, information is a key resource and examination a need in the associated circle of items (over their full life-cycle), creation resources and more (Manyika et al., 2015).

The Industrial Internet of Things is the greatest and most essential piece of the Internet of Things now, yet purchaser applications will have to make up for lost time from a spending point of view, for the most part from the beginning of 2018. All things considered, the Industrial Internet of Things is unmistakably increasingly imperative and developed in the general I.o.T. picture (Giffi et al., 2015).

i) **Industry 1.0 (1784).** The development of steam motors kickstarted Industry 1.0. The manufacturing was simply work-oriented and tiresome (Boyes et al., 2018).

ii) **Industry 2.0 (1870).** The principal sequential construction system generation was presented. This development was a major relief for specialists as their work was limited to a conceivable degree. Henry Ford, the father of large-scale manufacturing and the sequential construction system, presented the procedure in a vehicle fabricating plant by Ford to improve profitability by utilising the transport bell mechanism (Boyes et al., 2018).

iii) **Industry 3.0 (1969).** Included progression of electronic innovation and mechanical application on autonomous miniaturisation of the circuit board through programmable rationale controllers, Industry 3.0 applied autonomy to essentially robotise and automatically increase the production (Boyes et al., 2018).

iv) **Industry 4.0 (2010).** The vision of associated ventures through the internet was satisfied with the introduction of Industry 4.0. Smart gadgets communicate with one another and create valuable insights (Boyes et al., 2018).

Activities that advance the modernised ongoing changes are extending between researchers, adventures and planners of the "smart factory" (Market Research Future, 2018; Boyes et al., 2018).

In Industry 4.0, manufacturing structures and the articles they make are not just related, drawing information from the physical world into the mechanical space. Or maybe, Industry 4.0 takes this thought one step further (Kolias et al., 2017; Schneider, 2017): that information is then dismantled and used to drive further smart actions in the physical world, completing a physical-to-mechanised-to-physical cycle of action and instructed reaction. This cycle of astute, self-overseeing propelled development and the Industry 4.0 advancements that drive it impact the habits by which associations attract their customers and meet customers' reliably developing tendencies. Further, and perhaps most importantly, it enables producers to move their motivating force from things to constant, data-driven organisations (Kaufmann, 2015). Without a doubt, from early on, research and arrangements took account of the board and affiliate's trade service (Heppelmann & Porter, 2014), and related developments created opportunities to improve efficiency and overhaul customer experiences, helping creators attract and keep customers similarly to gigantic organisations' driven values (Daugherty et al., 2015). I.o.T. has a large number of uses in assembly plants. It can improve the life-cycle in an assembly plant, as I.o.T. gadgets normally screen progression cycles (McDevitt et al., 2014), and direct dispersion focuses similarly on inventories. It is one reason enthusiasm for I.o.T. devices has taken off over the last few decades (Rose et al., 2015). I.o.T. in a collaboration with transportation will rise to $40 billion value by 2020.

The foundation of I.I.o.T. is the way by which the advanced assets (the things, machines, goals and circumstances) can be related with business specialists and processes (Lukac, 2016). By 2023, one estimation calls for 20 billion related devices with sensors, actuators and embedded enrolling capacities. The framework is the basic establishment that reinforces particular application requirements and various sending conditions in the colossal extent of industry regions and related industry-unequivocal applications affected by I.I.o.T. (Baheti & Gill, 2011). Present-day framework organisation is a social event of advances at the Internet Protocol (IP) layer (Dorsemaine et al., 2016) and underneath that engage the difference in organisations. There are various choices in advancements, both existing and emerging (Köster et al., 2009). What are the framework organisation needs, what applications will help on the mechanical framework and what is the associated situation and conditions? These are key requests to answer when describing a strategy for a development decision and accomplishing a strong game plan solution (Fleisch et al., 2014).

3.1.2   I.o.T. MANUFACTURING OPERATIONS

3.1.2.1   Intelligent Manufacturing

As manufacturers migrate from conventional manufacturing plants to I.o.T.-related, IP-based structures, there is an increase in new vulnerabilities. Unavoidable in interfacing methods and segments of smart assembly is an improvement of the computerised ambush surface. Each reason for affiliation transforms into an extra risk of strikes and cybercrimes that can provoke impedance, remote access, theft of secured advancement and data corruption or alteration (Accenture, n.d.). Though many tried and true security instruments remain feasible, they are not always organised into systems from the earliest beginning stage. To ensure adequate security, security challenges have similarly blocked the pace of choice of new I.o.T. developments, progressive changes and plans for activity that could hugely improve frames, redesign power and bring new organisations to customers (Köster et al., 2008). Grievously, attempts that don't keep pace will feel that it is logically difficult to battle with their undeniably serious assistants who are dealing with the test head-on (Köster et al., 2009). Producers should work with experienced integrators, specialists and advancement assistants who have quite recently appeared and life length in partner, checking and adjusting sharp gathering structures. Experienced assistants can give the support expected to develop the best system to meet the business needs (Emmrich et al., 2015; Dorsemaine et al., 2016; Dalenogare et al., 2018). Well-being endeavours must be embedded into all production systems from the earliest starting point, engaging protection creation and security against computerised risks. There is no one-measure-fits-all course of action. Or perhaps makers must work with professionals to verify and watch the devices, the framework, the data and the item's courses of action and applications driving I.o.T.'s excellent production systems. Gemalto offers organisations and game plans that give apparent all-the-way security (Schneider, 2017), whatever the accessibility suggests, from cell to L.P.W.A.N. to fixed frameworks. The potential for sharp gathering is gigantic and should not be hampered by security questions. Industry pioneers must be confident while making or retrofitting smart assembling plants. It is essential that game plans be carefully and prudently picked and considered for extraordinary execution. Smart producing is empowered by I.o.T. associated gadgets, large amounts of information, examination of information, mechanical autonomy, sensor advances and artificial intelligence (Boyes et al., 2018; Jeschke et al., 2017; McDevitt et al., 2014). All of these innovations are utilised together to enhance fabricating forms, help producers and guard specialists. Here are some more potential good effects of smart manufacturing:

- **Greater operating efficiency.** The capacity to comprehend what's going on at each phase of the manufacturing procedure permits plant administrators to actualise ongoing arrangements and limit machine downtime. This prompts more prominent effectiveness and decreased costs (O'Halloran & Kvochko, 2015).
- **Minimal machine downtime.** The artificial intelligence algorithm can utilise activity and setback logs to transform ongoing I.o.T. sensor information into prescient maintenance knowledge. This enables human labourers to be aware of potential part problems, or even empower these segments to act naturally 'healing' (Lukac, 2016; O'Halloran & Kvochko, 2015).
- **Increased worker safety.** Computerised following and investigation of individuals' developments and movements inside a smart plant can help moderate dangers. Security frameworks caution people inside the industrial facility about explicit injury risks, looking for slips in the centre or mix-ups during new or routine undertakings. These frameworks could likewise control or send for restorative help if needed (Kaufmann, 2015).
- **Optimised inventorying.** Real-time sensor information can be utilised to follow the area of parts and items previously, both amid, and in the wake of, assembling. Examination gives the essential data to request and minimise costs on requests with consistent scheduling.
- **Supply chain management.** Manufacturers can utilise I.o.T. gadget geolocation to follow shipments, parts and items. This information would then be utilised to alter planning on parts and arrange and give bits of knowledge that allow items to effectively be reused, bringing down the expense of materials for manufacturing (Dorsemaine et al., 2016).

As per market research, the smart manufacturing plant advertising is anticipated to achieve 205.42 billion USD by 2022 (Wan et al., 2018), developing at a CAGR of 9.3% somewhere in the range of 2017 and 2022.

3.1.2.2   Asset Management

Smart manufacturing enables manufacturing plant administrators to naturally gather and investigate information to settle on better-educated choices and enhance production (Liu et al., 2018; Schneider, 2017). That information is broken down and joined with the relevant data and after that shared with approved partners. I.o.T. innovation, utilising both wired and remote networks, empowers this stream of information, giving the capacity to remotely screen and oversee procedures and change generation of designs rapidly and progressively when required to optimise processes (Emmrich et al., 2015).

The internet of things (I.o.T.) is turning into a quick embraced innovation arrangement which is basically because of the explosion of simple sensors, reasonable network, versatile cloud storage, ingestion, preparation and capacity for storing immense amounts of organised/unstructured information in huge information stages, universal portable applications with the help of AI tools (Rose et al., 2015). Technology organisations and specialist co-ops are progressively besieging organisations with creative arrangements and ways on how they can interface things to profoundly change their business activities.

I.o.T.-powered smart resource checking arrangements do everything that conventional arrangements do, like telling the associations where the benefit is, what is the state of the advantage, overseeing resource life-cycle, control forms, etc. (Hermann et al., 2015). Likewise, it adds knowledge to robotised work processes, continuous alarms, pieces of knowledge from information, dynamic edge control of advantages, prescient support, cross-area investigation and ongoing visibility (Valdes, 2017).

I.o.T. smart asset management solutions typically comprise the following:

- **Remote Asset Tracking**
- **Asset Health/Condition Monitoring**
- **Asset Lifecycle Management**
- **Asset Workflow Automation**
- **Predictive Asset Maintenance**.

I.o.T. arrangements associate machines with individuals and with procedures and frameworks more than ever. This encourages automation (Falco et al., 2004). The human intervention is required just for basic leadership instead of performing ordinary assignments, rule-based pre-set activities, estimating field information or gathering review logs for administrative compliance (Falco et al., 2002). The essential preferred standpoint of smart asset monitoring is mechanising the majority of this. Along these lines, it expands precision, decreases cost, improves process proficiency and dispenses with resistance. Physical checks, routine errands and occasional observation would all be able to be decreased radically and now be made dependent on the real condition and utilisation of the asset (Roman et al., 2011).

Generally, producers are very much aware of what the I.o.T. can accomplish for them, as far as substantial and additional more profound bits of knowledge into their activities and supply chains (Dalenogare et al., 2018; Boyes et al., 2018).

For the last-mentioned, resource-escalated enterprises like assembling are anticipating that the I.o.T. should drastically improve their advantage in executive practices

(Boyes et al., 2018). Another investigation from Forrester and SAP separated what businesses are relied upon to embrace and use I.o.T. capacities for their advantage in the executive and upkeep departments (Jeschke et al., 2017).

Furthermore, the scientists found that numerous associations presently can't seem to decide official responsibility for deployments. The I.o.T. is a cross-function activity, so it's hard to tie down precisely who ought to oversee it and which division the financial plan for it should originate from. Prior authoritative storehouses make it difficult to facilitate I.o.T. utilisation crosswise over practical regions, and if administrators aren't completely sold on the advantages, it might be hard to accumulate hierarchical support (Accenture, n.d.).

This is known as prescient support. Utilising a variety of implanted observing apparatuses, resources can reliably convey the condition of their warm properties, grease levels, oil testing, vibrations and other key markers of advantage health (Falco, 2004; Köster et al., 2008). This information helps resource supervisory crews spot shortages before they lead to unscheduled downtime. Here are a couple of reasons why this is so helpful to manufacturers:

- **Requires less human intercession at the floor level.** One of the significant issues fabricates are having today is finding skilled energetic workers to fill the shoes of those leaving (Poovendran, 2010). The I.o.T. motorises a lot of assessment and work demand filling, decreasing the requirement for large human groups (ITU, 2012).
- **Schedule support around generation runs.** Today, numerous producers are consolidating the lean collecting perspective into their work. This anticipates that they should grow uptime faithful quality to ensure that customer requirements can be fulfilled as quickly as is conceivable (Jeschke et al., 2017). Observant help makes it easy to design breaks for maintenance, and manufacturing runs can be planned around scheduled upkeep along these lines, keeping up an essential barrier to missed deadlines (Rose et al., 2015).
- **More effectively track costs.** A prescient upkeep program starts with an incredible central database that screens spare parts, work orders, asset conditions and other fundamental data expected to make financial decisions concerning the organisation of key resources (Dorsemaine et al., 2016; Heppelmann & Porter, 2014). It's amazing for any business to cut costs if it hasn't the faintest idea where they started from. Data-driven assets on board make it possible to involve resources for the most imperative need practices and legitimise further theories to officials (Poovendran, 2010; Boyer, 2016).

The I.o.T.'s place in assembly is everything except certain, and keep in mind that there are impressive obstacles that organisations must defeat to utilise it.

3.1.2.3 Planning

The biggest hurdle for some organisations with regard to the Industrial Internet of Things (I.I.o.T.) isn't only the notable network or security issues. Just as overwhelming, if not more so, is choosing where to begin with I.I.o.T. so that it works well for the company (Curran et al., 2017; Manyika et al., 2015).

The fourth Industrial Revolution, also called Industry 4.0, is going all out. Industry 4.0 extends the physical condition with the advanced by conveying mechanical developments like enhancements once found only in sci-fi films. With numerous industry reports distributing patterns, examination and forecasts for Industry 4.0 through mechanical I.o.T., the numbers all point toward one shared trait: hyper-development. By 2020, the modern assembling division predicts that spending on I.o.T. will increase to $890 billion, while I.I.o.T. will add $14.2 trillion to the world-wide economy overall (Jeschke et al., 2017).

The move to advance drives a structural move in the modern part, and a lion's share of business pioneers are trying to make progress in this new period of manufacturing (Schneider, 2017). This development, with its fast changes, has numerous producers scrambling to execute vital I.o.T. initiatives (Antonakakis et al., 2017).

Inside the system of this idea, another important and corresponding idea emerges, the supposed 'Internet of Things' idea (I.o.T.) (Boyes et al., 2018). The I.o.T. grants communication among individuals and things, and at its most astounding level of advancement, the connection among things and things. This connection involves acquiring data from a situation, and preparing as well to send this data to different things over the internet (Kaufmann, 2015). These different things receive it which, thus, can set up another connection with another thing (Dorsemaine et al., 2016). In this process, things can essentially: get and send data over the internet; process the data they get from their situation or which they get from different things; decide to alter their execution or the execution of the framework to which they belong (Antonakakis et al., 2017).

3.1.2.4 Monitoring

Numerous organisations are attempting to obtain this I.O.T. technology (Jeschke et al., 2017). The way that computers had been modified with all the data they expected to work, however, and could rather detect their general surroundings and in this manner connect with raw numbers without involving human interaction, had introduced another era of computational technology (Antonakakis et al., 2017). The Industrial Internet of Things (I.I.o.T.) has taken this business innovation considerably further and connected it on a much more amazing scale (Boyes et al., 2018). This has prompted developments like the smart manufacturing plant and prescient innovation. By furnishing mechanical machines with sensors and preparing workers the whole way across the supply and delivery chains with the instruments to screen and react to the yield from these sensors, organisations have started to streamline all business activities. The inconvenience is that they regularly communicate through restrictive conventions that usually use RS-232/422/485 sequential links. While these sequential conventions are proficient and were regularly composed for a particular application, a large number of these applications never included all day, every day checking back-and-forth over TCP/IP systems. So as to carry these gadgets into the connected factory, Industry 4.0, as well as I.I.o.T. paradigm (Asenjo & Maturana, 2018), an association's architects should initially guarantee that the gadgets can communicate with the other hardware on the processing plant floor (Liu et al., 2018). Showing this basic execution information continuously drives profitability and increment throughput. This idea isn't restricted to associating, imparting and observing inside a company. This idea can likewise be stretched out to incorporate the supply and circulation chain to exhibit a far-reaching perspective on the whole operation (Dalenogare et al., 2018). The presentation of the internet of things and services into the assembling condition introduces a fourth mechanical conflict: Industry 4.0 (Jeschke et al., 2017). The goal of this investigation was to dissect the phenomenon of the topic Industry 4.0 identified with I.O.T. through an efficient survey of the internet of the learning base (Huberman, 2016).

Security has customarily implied physical confinement of mechanical equipment and connected networks (Daugherty and Berthon, 2015). In the event that nothing is associated with the computer hardware, the risk of security breaches is genuinely low. Network-free offices are rare as more businesses keep on growing their I.T. systems into processing plant settings. As businesses grasp this new reality, security ought to be monitored through careful system arranging and utilisation of I.P. address best practices. Switches can be sent inside a system to restrain organised traffic to explicit kinds of traffic or to explicit clients, limiting the danger of a cyber-attack. Another strategy is the execution of N.A.T. (Network Address Translation) (Dalenogare et al., 2018). N.A.T. is a method that screens gadgets on a system from inbound access, however, it doesn't influence traffic on a system. The internet of things (I.o.T.) empowered assembling and cloud manufacturing. Likenesses and contrasts in these themes are featured depending on our examination. Next, we depict overall developments in smart assembling, including administrative key plans from various nations and vital plans from real global organisations in the European Union, United States, Japan and China (Yaqoob et al., 2017).

I.o.T. is a standout amongst the most discussed advancements in the innovation showcase. Numerous organisations are endeavouring to embrace this innovation. Likewise, numerous new organisations are endeavouring to enter in the race of creating great I.o.T. applications and I.o.T.-upheld frameworks.

Condition observing is only checking the state of any framework or machine. By observing the machines consistently, it turns out to be anything but difficult to see whether the machines need fixing or not (Dorsemaine et al., 2016).

The machine condition checking report offers information that predicts apparatus failure. The condition observation tracks changes in the temperature of the machine, regardless of whether it is vibrating, and so forth. It additionally checks the yield of machines to recognise lop-sidedness, consumption, wear, misalignment, silt development, or ineffectively lubricated parts (Daugherty et al., 2015).

3.1.2.5 Types of Condition Monitoring

i) **Route-Based Monitoring.** In course-based checking, the specialists record information continuously. It carries out a deep well-being investigation of the machine. In the wake of breaking down the information, specialists choose if there is a requirement for cutting-edge examination (Kargermann et al., 2013).

ii) **Versatile Machine Diagnostics.** This sort of condition checking requires a compact machine that checks the sensors fitted in the machines to gather the information (Kargermann et al., 2013).

iii) **Industrial Facility Assurance Test.** Asset machines also need condition checking. This procedure is called item testing (Geisbauer et al., 2016). It decides conceivable failure methods of the gadget.

iv) **Online Machine Monitoring.** Online machine observing checks the machine as it runs and works. The report is sent to the crucial server. Here, data is analysed and support is booked (Boyes et al., 2018).

As indicated by a report, 82% of affiliations have faced an unconstrained stoppage throughout the latest three years. An unconstrained stoppage is a significant issue as it can cost the company US$260,000 for every hour. As showed by another investigation, 64% of unconstrained stoppages happened due to machine failure or apparatus breakdown.

There can be different purposes for the breakdown of a machine. It can happen in light of maltreatment of the machine, or due to inappropriate servicing or non-attendance of viable machines' condition following.

Companies endeavour to keep ahead of failures to keep work progress going. Breakdowns of machines and rigging can cause a disaster to the business. The web of things has exhibited the best results for condition checking (Edmondson, 2017; Wan et al., 2018).

As of late, Mercedes-Benz developed a truck mechanical production system for Daimler Trucks, its effective advertisement vehicles-maker, utilising the Industry 4.0 ideas of hyperconnectivity. The generation lines comprised various sensors, particularly light sensor innovation in forklifts to encourage the materials stream. With a 15% improvement in assembling effectiveness, Mercedes-Benz has chosen to broaden the ideas of Industry 4.0 to all its creation processes for material assemblies. The sensors would be a tremendous piece of this provident usage by Mercedes-Benz (Boyes et al., 2018; Sisinni et al., 2018).

v) **Machine–Human Interaction.** Industry 4.0 will be accompanied by changing assignments and requests for the human in the industrial facility. As the most adaptable element in the cyber-physical production system, workers will be tasked with a huge assortment of occupations going from detail and observing to checking manufacturing techniques.

The utilisation of built-up collaboration innovations and allegories from the purchaser merchandise showcase is by all accounts promising. This chapter shows answers for the mechanical help of specialists, which actualise the portrayal of a digital physical world and the collaborations happening as insightful U.I.s. Other than innovative methods, the chapter calls attention to the prerequisites for sufficient capability methodologies, which will make them required, between disciplinary comprehensions for Industry 4.0 (Jeschke et al., 2017). Various levels of engineering of the smart processing plant were proposed first, and after that the key advancements were broken down from the parts of the physical asset layer, the system layer and the information application layer (Antonakakis et al., 2017). Furthermore, we talked about the serious issues and potential answers for key developing advances, for example, the internet of things (I.o.T.), enormous information and distributed computing, which are installed in the assembly procedure. In the usage of a smart production line, the I.I.o.T. is utilised to coordinate the hidden hardware assets. Likewise, the assembling framework has capacities for observation, interconnection and information integration (Yaqoob et al., 2017). The information investigation and logical choice are utilised to accomplish manufacturing contracts, equipment administration and quality control of items in the smart plant. Further, the internet of administrations is acquainted with virtualising the assembling assets from a neighbourhood database to the cloud server. Through the human–machine communication, the worldwide collective procedure of smart manufacturing focused on a request-driven market is created. As the most adaptable element in digital physical generation frameworks, workers will be tasked with an extensive assortment of occupations going from individual checking to confirmation of manufacturing techniques. Through mechanical help, it is ensured that workers can understand their maximum capacity and receive the job of vital leaders and adaptable issue-solvers (Püschel et al., 2016). The utilisation of set-up collaboration innovations and similarities from the buyer merchandise showcase is by all accounts promising. The mechanical help of specialists, which execute the portrayal of a digital physical world and the collaborations happening as smart client interfaces (Albert, 2015). Other than innovative methods, the chapter discusses the necessity for satisfactory capability procedures, which will make the required, between disciplinary comprehensions for Industry 4.0. adaptability for assembling, mass tended to by the umbrella terms 'Industry 4.0' and 'Advanced Manufacturing', and also briefly consider a few difficulties or obstacles to improving profitability in the workplace. Among them, work division between human specialists and smart machines in innovation-rich workplace brings up issues about the idea of human A.I. in the manufacturing plants of the future. This adjusts the change in outlook in work-based and professional instruction and related didactical ideas. An ongoing study has uncovered that just 13% of labourers in O.E.C.D. nations and economies utilise key data handling abilities, in particular education, numeracy and critical thinking aptitudes once a day with higher capability than P.C.s (Asenjo & Maturana, 2018). Psychological registering tries to replicate human abilities through structure artificial models and calculable calculations sent for taking care of human sorts of issues (undertakings) and exchanging human basic leadership procedures to smart machines (Wan et al., 2018; Liu et al., 2018). Human learning has been considered a subject in the field of training, teaching methods and psychology in connection to the learning speculations (behaviourism, cognitivism, constructivism and humanism), learning styles, academic models, idea learning and instructive neuropsychology (Boyes et al., 2018). This has prompted a wide variety of meaning of human learning and accordingly, an all-inclusive agreement on any single definition is non-existent (Boyes et al., 2018). Learning is a change in conduct, or in the ability to carry on in a given manner. Constructivist speculations, for example, social constructivism, arranged learning and connectivism (Edmondson, 2017) have set up the 'establishment for most of instructing strategies that have grabbed hold as of late' (Boyes et al., 2018). The five-classification model of adult learning, in which the experience and expertise dimension of a student are profoundly connected, considers learning with making ideas and significance as a matter of fact (United Nations Educational Scientific and Cultural Organization, 2017). Human and machine cooperation and joint effort lay the ground for hybridisation of the learning ideas, in which 'shared learning' happens. This is likewise influenced by various potential limits for human and machine in performing distinctive assignments, for example, mechanical employment and basic leadership. Quality and execution variety in completing an errand are the key markers.

vi) **Cyber-Physical System.** During the time spent in data trade (correspondence), computational procedures are quite often included. At first, the way toward measuring it is still straightforward; however, as it advances, the necessities of the undeniably quick procedure figuring process at that point turns out to be progressively mind-boggling. Something associated with the procedure of complex figuring incorporates arranging a specific measure of information into data. Computational speed is indivisible from the innovative advances that happen in it. Advances in innovation likewise get changed measurements of gadgets, sensors and actuators. A framework that consolidates the capacities of processing, correspondences and information stockpiling, so as to screen or control the substances that exist in the physical world is called a cyber-physical system (C.P.S.) (Wan et al., 2018). Just as the internet is changing how people convey information, how and where to get to data and make the way toward purchasing and selling, so the C.P.S. will change the way people connect with and control the physical world around them. Cyber-physical systems is a term created right now in the workshop composed by Professor Raj Rajkumar of Carnegie Mellon University, PA, U.S.A. In the introductory materials at the opening of the workshop, it is expressed that the framework that consolidates the capacities of processing, interchanges and information stockpiling, so as to screen or control the elements that exist in the physical world is called cyber-physical systems (Wan et al., 2018).

Stringent confinements are connected to the C.P.S., which is the control that ought to be done in a nearby circle state with a specific time limit. The running procedure is relied upon to speak to occasions that happen in the physical world, which are parallel occasions. This is a test in C.P.S. inquiries on account of the computational (digital) process are sequential (MRS Research Group, 2017). It is calculated that C.P.S. themes will deliver new science within 10–20 years, and one of the discussions is the way to structure the incorporation among physic and digital all the more normally.

3.1.3 IMPORTANCE OF DATA IN I.O.T. MANUFACTURING

I.o.T. sensors assume a vital job in manufacturing. The majority of the information gathered by I.o.T. gadgets gives a significant understanding of assembling execution which is essential on a bigger scale for worldwide client satisfaction (Baheti & Gill, 2011).

At the point when makers choose to apply an internet of things approach, they additionally settle on a decision to upgrade their administration and give better items, administrations and better quality. All things considered, as innovation improves so does machine learning. All things considered, a superior and quicker generation process with expanded consumer loyalty just prompts more prominent profitability, by and large.

In reality, I.o.T. investigation found that assembling by smart gadgets prompts an income increase of 33.1%, better quality for at any rate 11% and 22% expanded consumer loyalty. What's more, among 600 assembling organisations, 97% concur that the I.O.T. is the biggest innovation, while 83% of producers as of now have these gadgets installed (Valdes, 2017).

Assembling clearly covers numerous kinds of items, tasks, forms and an immense variety of exercises, segments, machines, individuals, accomplices, data frameworks, etc. It is far from crude materials to completed merchandise and it is unavoidably related to supply chains, coordination and transportation also.

In the event that we take a look at assembling as a mechanical creation in any of the phases where crude materials are transformed into items or items are utilised to construct different items, we unmistakably see a colossal market that is exceptionally interconnected (Giffi et al., 2015).

3.1.4   BENEFITS OF I.O.T. IN MANUFACTURING

At this point, the advantages of I.o.T. are very clear. Be that as it may, the assistance of I.o.T. comes to over and above these incredibly effective assembling procedures and control. Hence, the aggregate points of interest include (Boyes et al., 2018):

i) **Enables associations among makers and machines.** Modern systems administration is an accumulation of advances at the Internet Protocol (I.P.) layer and beneath that empowering the changes of enterprises. There are numerous decisions in innovations, both existing and developing. What are the systems administration needs, what applications will help the mechanical system and what are the arrangement circumstances and conditions? These are key questions to answer when characterising a procedure for innovation choice and achieving a solid sending arrangement. Mechanical systems administration is not quite the same as systems administration for the endeavour or systems administration for purchasers. First, there is the union of Information Technology (I.T.) 4 and Operational Technology (O.T.). Imperative systems administration considerations incorporate whether to utilise wired or remote, how to help portability (for example, vehicles, hardware, robots and labourers) and how to reconfigure segments of industrial systems to improve profitability and execution by conveying fundamental operational data (Edmondson, 2017; Yaqoob et al., 2017). For instance, condition checking and vitality observing gather information from each bit of equipment in the activity that can be utilised to make profiles for each bit of hardware, such as control use over a hardware cycle and vibration profiles, among typical tasks. This information is frequently transmitted to the cloud for investigation and, when contrasted with past profiles, can add to prescient choices about when to support the equipment, along these lines pre-empting hardware failure during production. There are a lot more developing situations that influence these layers and that should be bolstered. We offer a few models here. They are not intended to be comprehensive (Edmondson, 2017; Yaqoob et al., 2017).

ii) **Control over joint frameworks**. In the field of observation and control, control gadgets with inherent P.C.s called programmable rationale controllers and procedure control stations are circulated close to the equipment they are proposed to control straightforwardly. Associating these gadgets to a control neighbourhood (LAN) makes it possible to gather information in a focal checking and task gadget for joint observation and operation (Jeschke et al., 2017). This incorporates not just use as checking and control frameworks for individual offices, for example, water or sewage treatment plants and siphoning stations, but in addition the utilisation of telemetry or devoted Internet Protocol (I.P.) lines to empower the interconnection and locale-wide administration of such offices over the region served by a utility. Along these lines, the observation and control frameworks themselves can shape a system of things.

iii) **Deeper understanding of the assembling procedure**. The web of things needs no more introduction. Right when these I.o.T. limits are completed in the industrial and manufacturing space, it winds up in industrial I.o.T. (Emmrich et al., 2015). This development is an amalgamation of different advances like A.I., Big Data, sensor data, M.2.M. correspondence and computerisation that have existed in the cutting-edge view for quite a while.

Present-day internet makes a related endeavour by merging the information and operational divisions of the business, thus improving detectable quality, boosting operational capability, and assembly productivity and reducing the multi-faceted idea of methodology in the business. Mechanical I.o.T. is a transformative collecting method that improves quality, prosperity, proficiency in an industry (Dorsemaine et al., 2016; Hartmann & Halecker, 2015; Schneider, 2017).

iv) **Data examination settles on more brilliant business choices.** This flood of information is filling across the board I.o.T. selections as there will be about 30.73 billion I.o.T.-associated gadgets by 2020 (Jeschke et al., 2017). Data analytics (D.A.) is characterised as a procedure, which is utilised to look at all shapes and sizes of informational collections with differing data properties to remove important ends and significant bits of knowledge. I.o.T. information comes in large volumes, is exceedingly unstructured and varies as far as range (content, picture or recordings). In addition, while the operational innovation identifies with the information gathered from temperature sensors, weight sensors, tablets, smart assembling gadgets/devices and so on, the data innovation identifies with the information gathered from big business frameworks, inheritance frameworks, E.R.P., C.R.M. and fund frameworks. Taking a look at just the O.T. or I.T. information in storehouses won't give the vital outcomes. The O.T. and I.T. information must be consolidated to have good potential. Lamentably, customary examination instruments and advances are intended to look at just the I.T. information and don't work straightforwardly on this consolidated dataset (Yaqoob et al., 2017).

v) **Enhanced generation work process via automated and advanced creation forms.** In the A.P.e.J. (Asia Pacific, excluding Japan) area, near 33% of all I.o.T. spending (equipment, programming, administrations and network consolidated) will be for the manufacturing business in 2020 (Heppelmann and Porter, 2014). Also, in different areas, producing positions are first, yet with slower pieces of the pie of complete I.o.T. spending. In the U.S., for example, I.o.T. spend by the assembling business will represent around 15% of all-out I.o.T. purchases.

vi) **Devices associated with the system permit hardware the executives can control from any area.** The industrial internet of things (I.I.o.T.) utilises both production network of the board and smart coordination. Be that as it may, it demonstrates its genuine advantages in prescient upkeep and resource as follows: a blend of temperature, H.V.A.C. and synthetic sensors controls the states of workplaces. The area following R.F.I.D. labels distinguishes pivotal resources and empowers an unheard of dimension of transportation computerisation in smart factories (Kolias et al., 2017; Falco et al., 2004; Accenture, n.d.). Eventually I.o.T. executions improve the manufacturing plant's operational effectiveness and spare time and expenses. One case of how I.o.T. can improve producing forms is with computerised twins. An advanced twin is a virtual portrayal of a physical article, which can be an individual thing, a machine or an entire assembly plant. With the assistance of advanced twins, the creation line can be checked carefully and continuously to distinguish defects when possible (Schneider, 2017).

vii) **Sensor information gives a superior outline of intensity utilisation.** The industrial internet of things (I.I.o.T.) is frequently alluded to as Industry 4.0 to signify the fourth modern upheaval, that of interfacing items, machines, administrations and people through the cloud (Lukac, 2016). I.I.o.T. is pushing toward a time of expanded interconnectivity, which is estimated before long to exceed shopper I.o.T. (Hartmann and Halecker, 2015). This new time of smart assembling requires new biological communities to encourage full manufacturing plant computerisation and continuous observation with the point of expanding efficiency, empowering prescient upkeep and upgrading the production network and resource detectability while building up a more secure and progressively safer condition. Interconnection of machines requires the organisation of sensor hubs, gadgets that take estimations (for instance, the area of an important resource, the ecological conditions in which a food item is being put away or the warmth produced by a damaged siphon), which can be put away or transmitted to a centre point for further preparing. Many sensors are currently cabled, which is normally unrealistic when bits of equipment may be moved around and full re-wiring is then required. Retrofitting maturing creation lines with sensors can be dubious, if not absolutely outlandish, when information should be obtained on moving parts or close hot machines. In this way, as a rule, making the gadget self-ruling as far as the way it is controlled, for instance, by utilising batteries, makes a great deal of sense (Falco et al., 2004; Antonakakis et al., 2017). One favourable position of batteries is their moderately minimal effort. However, what is imperative is the all-out expense of responsibility for the gadget. Obviously, coin cells are presently vigorously commoditised and can be bought for around a dollar typically. The genuine expense goes up essentially, in any case, when you include the expenses to pay a specialist to change a failing battery. In a production line that may have hundreds or even a great many sensors introduced, this could end up being somebody's work all day. In any case, past establishment costs, what is significantly increasingly exorbitant to the organisation is the period when the battery is going to fall flat or has failed. When this has been seen and the battery has been changed, key information may have been lost. In actuality, most batteries just last two to five years, which is excessively short contrasted with the normal existence of the gadget they control, which is maybe 10 to 15 years (Jeschke et al., 2017). In this way, a method for reducing fuelling and gadgets 'interminably', that can obtain information and send them remotely to an information centre point, would be perfect.

viii) **Time-sparing and cost anticipation.** This pattern will proceed for quite a long time to come, especially with the developing ubiquity of cell phones, everything being equal. Tending to control utilisation for battery-fuelled dependably on I.o.T./I.I.o.T. gadgets which depend on many electronic segments, including sensors is basic to their business success. The interest in ultra-low-control sensors has sped up the race to crush each and every mW from components (Ahmed et al., 2016).

3.1.5 CONCLUSION

The industrial internet of things is an arrangement of physical things, structures, stages and applications that contain embedded development to grant and communicate information to each other, the external condition and with people. The allotment of industrial I.o.T. is being engaged by the improved availability and sensibility of sensors, processors and various advancements that have empowered receipt of and access to steady information. The industrial internet of things and Industry 4.0 are still 'fluffy'. Since these ideas have no exact and generally acknowledged definitions, we exhibited some viewed as important by logical writing. I.I.o.T. can give imperative advantages to organisations in a wide range of enterprises. For instance, other than concentrating just on the specialised issues, the connections between all partners, from people to associations, and from organisations to governments, should likewise be thought about. Over the coming years, businesses and governments must strengthen their endeavours and heighten speculations, yet additionally should change their ways of dealing with training, abilities and work.

REFERENCES

Accenture. n.d. "Industrie 4.0 – Die Nächste Industrielle Revolution." https://www.accenture.com/de-de/service-industrie-4-0-die-nachste-industrielle-revolution.

Ahmed, Ejaz, Ibrar Yaqoob, Abdullah Gani, Muhammad Imran, and Mohsen Guizani. 2016. "Internet-of-Things-Based Smart Environments: State of the Art, Taxonomy, and Open Research Challenges." *IEEE Wireless Communications* 23 (5): 10–16. doi:10.1109/MWC.2016.7721736.

Albert, Mark. 2015. "Seven Things to Know About the Internet of Things and Industry 4.0." *Modern Machine Shop* 88 (4): 74–81. http://www.redi-bw.de/db/ebsco.php/searc h.ebscohost.com/login.aspx?direct=true&db=buh&AN=109184251&site=ehost-live %5Cnhttp://content.ebscohost.com/ContentServer.asp?T=P&P=AN&K=109184251 &S=R&D=buh&EbscoContent=dGJyMNHr7ESepq84zOX0OLCmr02ep7FSsa24TL CWxWXS&C.

Antonakakis, Manos, Tim April, Michael Bailey, Matthew Bernhard, Ann Arbor, Elie Bursztein, Jaime Cochran, et al. 2017. "Understanding the Mirai Botnet." *USENIX Security*: 1093–1110. doi:10.1016/j.religion.2008.12.001.

Asenjo, Juan L., and Francisco P. Maturana. 2018. Industrial Internet of Things Data Pipeline for a Data Lake. US15/199,869, issued 2018. https://patents.google.com/patent/US1013 5705B2/en?oq=inventor:(Juan+L.+Asenjo)+15%2F199%2C869.

Baheti, R., and H. Gill. 2011. "Cyber-Physical Systems." *The Impact of Control Technology* 12 (1): 161–166.

Boyer, Stuart A. 2016. *SCADA: Supervisory Control and Data Acquisition, Fourth Edition.* International Society of Automation.

Boyes, Hugh, Bil Hallaq, Joe Cunningham, and Tim Watson. 2018. "The Industrial Internet of Things (IIoT): An Analysis Framework." *Computers in Industry* 101: 1–12. doi:10.1016/j.compind.2018.04.015.

Curran, Chris, Dan Garrett, and Tom Puthiyamadam. 2017. "A Decade of Digital Keeping Pace with Transformation." Digital IQ: PWC. 2017. https://www.google.de/url?sa=t &rct=j&q=&esrc=s&source=web&cd=1&cad=rja&uact=8&ved=0ahUKEwiOzJ Pixb_UAhUFshQKHSFAB-MQFggoMAA&url=https%3A%2F%2Fwww.pwc.com%2 Fus%2Fen%2Fadvisory-services%2Fdigital-iq%2Fassets%2Fpwc-digital-iq-report.pdf &usg=AFQjCNFcn1L0TSy5kn.

Dalenogare, Lucas Santos, Guilherme Brittes Benitez, Néstor Fabián Ayala, and Alejandro Germán Frank. 2018. "The Expected Contribution of Industry 4.0 Technologies for Industrial Performance." *International Journal of Production Economics* 204: 383–394. doi:10.1016/j.ijpe.2018.08.019.

Daugherty, Paul, Prith Banerjee, Walid Negm, and Allan E Alter. 2015. *Driving Unconventional Growth through the Industrial Internet of Things.* Accenture. doi:10.1016/j.molcata.2003.10.018.

Daugherty, Paul, and Bruno Berthon. 2015. "Winning with the Industrial Internet of Things: How to Accelerate the Journey to Productivity and Growth." *Accenture.* https://ww w.accenture.com/t00010101t000000z__w:_/it-it/_acnmedia/pdf-5/accenture-industrial -internet-of-things-positioning-paper-report-2015.pdf.

Dorsemaine, Bruno, Jean Philippe Gaulier, Jean Philippe Wary, Nizar Kheir, and Pascal Urien. 2016. "Internet of Things: A Definition and Taxonomy." In *Proceedings - NGMAST 2015: The 9th International Conference on Next Generation Mobile Applications, Services and Technologies*: 72–77. doi:10.1109/NGMAST.2015.71.

Edmondson, M., and A. Ward. 2017. "Tackling the Disconnect between Universities, Small Businesses and Graduates in Cities and Regions." *Gradcore.* https://www.eurashe.eu/li brary/mission-phe/EURASHE_AC_LeHavre_170330-31_pres_EDMONDSON-WAR D.pdf.

Emmrich, Volkhard, Mathias Döbele, Thomas Bauernhansl, Dominik Paulus-Rohmer, Anja Schatz, and Markus Weskamp. 2015. *Geschäftsmodell-Innovation Durch Industrie 4.0.* Dr. Wieselhuber & Partner GmbH Und Fraunhofer IPA. https://www.wieselhuber .de/migrate/attachments/Geschaeftsmodell_Industrie40-Studie_Wieselhuber.pdf.

Falco, J., K. Stouffer, A. Wavering, and F. Proctor. 2002. *IT Security for Industrial Control Systems.* US Department of Commerce, National Institute of Standards and Technology.

Falco, Joe, James Gilsinn, and Keith Stouffer. 2004. "IT Security for Industrial Control Systems: Requirements Specification and Performance Testing." In *Proceedings of the Homeland Security Symposium & Exhibition.* http://citeseerx.ist.psu.edu/viewdoc/dow nload?doi=10.1.1.75.4232&rep=rep1&type=pdf.

Fleisch, Elgar, Markus Weinberger, and Felix Wortmann. 2014. "Geschäftsmodelle Im Internet Der Dinge." *HMD Praxis Der Wirtschaftsinformatik* 51 (6): 812–826. https:// doi.org/10.1365/s40702-014-0083-3.

Geisbauer, Reinhard, Jesper Vedso, and Stefan Schrauf. 2016. *Industry 4.0: Building the Digital Enterprise.* 2016 Global Industry 4.0 Survey. doi:10.1080/01969722.2015.100 7734.

Giffi, Craig, Ben Dollar, Michelle Drew, Jennifer McNelly, Gardner Carrick, and Bharath Gangula. 2015. *The Skills Gap in U.S. Manufacturing, 2015 and Beyond.* Deloitte Development LLC. http://www2.deloitte.com/content/dam/Deloitte/us/Documents/ manufacturing/us-pip-the-manufacturing-institute-and-deloitte-skills-gap-in-manuf acturing-study.pdf.

Hartmann, Matthias, and Bastian Halecker. 2015. "Management of Innovation in the Industrial Internet of Things." In *XXVI ISPIM Conference – Shaping the Frontiers of Innovation Management*, Hungary: 1–17.

Heppelmann, James E., and Michael Porter. 2014. "How Smart, Connected Products Are Transforming Competition." *Harvard Business Review* 11 (November): 64–88.

Hermann, Mario, Tobias Pentek, and Boris Otto. 2015. "Design Principles for Industrie 4.0 Scenarios: A Literature Review." *Technische Universitat Dortmund* 1 (1): 4–16. doi:10.1109/HICSS.2016.488.

Huberman, Bernardo A. 2016. "Ensuring Trust and Security in the Industrial IoT." *Ubiquity* 2016 (January): 1–7. doi:10.1145/2822883.

ITU. 2012. "Internet of Things Global Standards Initiative." *Internet of Things Global Standards Initiative.* http://www.itu.int/en/ITU-T/gsi/iot/Pages/default.aspx.

Jeschke, Sabina, Christian Brecher, Tobias Meisen, Denis Özdemir, and Tim Eschert. 2017. "Industrial Internet of Things and Cyber Manufacturing Systems." *Industrial Internet of Things*: 3–19. doi:10.1007/978-3-319-42559-7_1.

Kargermann, Henning, Wolfgang Wahlster, and Johannes Helbig. 2013. *Deutschlands Zukunft Als Produktionsstandort Sichern: Umsetzungsempfehlungen Für Das Zukunftsprojekt Industrie 4.0: Abschlussbericht Des Arbeitskreises Industrie 4.0.* https://www.bmbf .de/files/Umsetzungsempfehlungen_Industrie4_0.pdf.

Kaufmann, Timothy. 2015. "Geschäftsmodelle in Industrie 4.0 Und Dem Internet Der Dinge." *Journal of Chemical Information and Modeling* 53. Springer Vieweg. doi:10.1007/9783658102722.

Kolias, Constantinos, Georgios Kambourakis, Angelos Stavrou, and Jeffrey Voas. 2017. "DDoS in the IoT: Mirai and Other Botnets." *Computer* 50 (7): 80–84. doi:10.1109/ MC.2017.201.

Köster, Friedrich, Michael Klaas, Hanh Quyen Nguyen, Markus Brändle, Sebastian Obermeier, and Walter Brenner. 2009. "Collaboration in Security Assessments for Critical Infrastructures." In *2009 4th International Conference on Critical Infrastructures, CRIS 2009.* doi:0.1109/CRIS.2009.5071499.

Köster, Friedrich, Hanh Quyen Nguyen, Markus Brändle, Martin Naedele, Michael Klaas, and Walter Brenner. 2008. "ESSAM: A Method for Security Assessments by Embedded Systems Manufacturers." In *The 3rd International Workshop on Critical Information Infrastructures Security.* http://www.researchgate.net/publication/44939 453_ESSAM_A_Method_for_Security_Assessments_by_Embedded_Systems_M anufacturers/file/d912f507651794f09d.pdf.

Liu, Chih-Hao, Chirag Patel, Srinivas Yerramalli, and Tamer Kadous. 2018. "Unlicensed Spectrum Coverage Enhancement for Industrial Internet of Things." US15/697,212, issued 2018. https://patents.google.com/patent/US20180123859A1/en?oq=us15%2F69 7%2C212.

Lukac, Dusko. 2016. "The Fourth ICT-Based Industrial Revolution 'Industry 4.0' - HMI and the Case of CAE/CAD Innovation with EPLAN P8." In 2015 *23rd Telecommunications Forum, TELFOR 2015*, 835–838. doi:10.1109/TELFOR.2015.7377595.

Manyika, James, Michael Chui, Peter Bisson, Jonathan Woetzel, Richard Dobbs, Jacques Bughin, and Dan Aharon. 2015. *The Internet of Things: Mapping the Value beyond the Hype*. McKinsey Global Institute. https://www.mckinsey.com/~/media/McKinsey /Industries/Technology Media and Telecommunications/High Tech/Our Insights/The Internet of Things The value of digitizing the physical world/Unlocking_the_poten tial_of_the_Internet_of_Things_Executive_summary.ashx.

Market Research Future. 2018. "Industrial Internet of Things Market Analysis 2018 To 2022."

McDevitt, Valerie Landrio, Joelle Mendez-Hinds, David Winwood, Vinit Nijhawan, Todd Sherer, John F. Ritter, and Paul R. Sanberg. 2014. "More Than Money: The Exponential Impact of Academic Technology Transfer." *Technology & Innovation* 16 (1): 75–84. doi :10.3727/194982414x13971392823479.

MRS Research Group. 2017. "Global Industrial Internet of Things (IIoT) Market 2017 - Production, Sales, Supply, Demand, Analysis & Forecast to 2021." http://www.mrsr esearchgroup.com/report/108389#report-details.

O'Halloran, Derek, and Elena Kvochko. 2015. "Industrial Internet of Things : Unleashing the Potential of Connected Products and Services." In *World Economic Forum*, 40.

Poovendran, Radha. 2010. "Cyber-Physical Systems: Close Encounters between Two Parallel Worlds." *Proceedings of the IEEE* 98:1363–1366. doi:10.1109/JPROC.2010.2050377.

Püschel, Louis, Maximilian Röglinger, and Helen Schlott. 2016. "What's in a Smart Thing? Development of a Multi-Layer Taxonomy." In *2016 International Conference on Information Systems, ICIS 2016*.

Roman, Rodrigo, Pablo Najera, and Javier Lopez. 2011. "Securing the Internet of Things." *IEEE Computer* 44 (9): 51–58. doi:10.1109/MC.2011.291%0D.

Rose, K., S. Eldridge, and L. Chapin. 2015. *The Internet of Things: An Overview. The Internet Society*. https://www.internetsociety.org/wp-content/uploads/2017/08/ISOC-IoT-Ov erview-20151221-en.pdf.

Schneider, Stan. 2017. "The Industrial Internet of Things (IIoT): Applications and Taxonomy." *Internet of Things and Data Analytics Handbook*, 41–81. doi:10.1002/9781119173601.ch3.

Sisinni, Emiliano, Abusayeed Saifullah, Song Han, Ulf Jennehag, and Mikael Gidlund. 2018. "Industrial Internet of Things: Challenges, Opportunities, and Directions." *IEEE Transactions on Industrial Informatics* 14 (11): 4724–4734. doi:10.1109/TII.2018.2852491.

United Nations Educational Scientific and Cultural Organization. 2017. *Working Group on Education : Digital Skills for Life and Work. Broadband Commission For Sustainable Development*. https://unesdoc.unesco.org/ark:/48223/pf0000259013.

Valdes, Yann Ménière, and Ilja Rudyk; Javier. 2017. "Patents and the Fourth Industrial Revolution (EPO)." European Patent Office, Handelsblatt Research Institute. https:// www.lemoci.com/wp-content/uploads/2017/12/Patents-and-the-Fourth-industrial-R evolution-2017.pdf.

Wan, Jiafu, Chin Feng Lai, Houbing Song, Muhammad Imran, and Dongyao Jia. 2018. "Software-Defined Industrial Internet of Things, Wireless Communications and Mobile Computing."

Yaqoob, Ibrar, Ejchiaz Ahmed, Ibrahim Abaker Targio Hashem, Abdelmuttlib Ibrahim Abdalla Ahmed, Abdullah Gani, Muhammad Imran, and Mohsen Guizani. 2017. "Internet of Things Architecture: Recent Advances, Taxonomy, Requirements, and Open Challenges." *IEEE Wireless Communications* 24 (3): 10–16. doi:10.1109/ MWC.2017.1600421.

3.2   BIG DATA AND ITS IMPORTANCE IN MANUFACTURING

Deepak Mathivathanan and Sivakumar K.

3.2.1   INTRODUCTION

Big Data has become an unavoidable term both with academicians and practitioners in the present information age. Since the beginning of the 21st century, technology has been playing a major role in our lives with remarkable innovations in terms of digital devices which can be used to learn about human behaviour. In 2010, Eric Schmidt, the former C.E.O. of Google, reported that 'There were five Exabytes of information created between the dawn of civilisation through 2003, but that much information is now created every two days, and the pace is increasing' (Schmidt & Cohen, 2013). Most companies are storing and utilising enormous amounts of data as information for analysing their business's progress. IBM reported that every day 2.5 quintillion (2.5×1018) bytes of data is generated from various sources like in messages, digital pictures, invoices, videos, sensors, social media posts and from numerous other digital sources (IBM, 2014). By 2020, approximately 100 billion connected devices will be producing data and thus the application of Big Data analytics has become a necessity (Walport, 2014). This enormous amount of data is referred to as 'Big Data', and it can be systematically processed and analysed to understand the current trends towards developing competitive business models. Modern manufacturing facilities include technologies such as the Internet of Things (I.o.T.s) and cyber-physical systems (C.P.S.) which are capable of recording and transmitting raw low-level granular data. These captured data can be subjected to analytics and any modelling applications to derive insights to improve the existing operations. The analysis of the captured Big Data is called Big Data Analytics. Beyond the rhetoric, this chapter is dedicated to details about what are the difficulties encountered by the manufacturing industries and how Big Data can impact manufacturing in facing the challenges.

3.2.2   CHALLENGES IN MANUFACTURING INDUSTRIES

Manufacturing sectors have always been highly competitive and innovative technologies in the last decade have forced them to change to streamline their production and their traditional methods. With the introduction of information technology, manufacturing companies generate massive volumes of data, but only a few make use of it. Hence, the manufacturing sector has great potential to grab the opportunities Big Data can offer. Like any other sector, the manufacturing sector also handles many challenges in the current information age. This section intends to highlight the various challenges in manufacturing industries.

   i) **Growing sustainability issues.** Manufacturing industries are the ones dealing with major environmental issues of the planet and they involve a lot of material and human resources. Hence, the industry faces more sustainability issues than other industries. They are considered a major contributor towards climate change and are also handling more social issues due to difficult working conditions. Unless the functioning of the industry improves drastically, the adverse environmental and social issues will remain a major barrier for sustainable development.

ii) **Stringent regulatory requirements.** With customers and society having increasing awareness of green and sustainability issues, the governments have taken stringent regulatory measures. In the last decade, regulations have been imposed by governments to protect nature and to ensure safe working environments. Hence, manufacturing industries are facing major challenges in complying with regulatory standards.

iii) **Adopting to latest technologies.** Most of the manufacturing industries used human labour to a greater extent in their production and plant operations traditionally. With recent advances in technology and with Industry 4.0 changing the way industries operate all over the world, industries find it more difficult to adopt emerging technologies. Moreover, insufficient knowledge and fear for change are important obstacles to taking up innovative technologies like cloud computing, I.o.T.s, etc.

iv) **Need for supply chain visibility.** Supply chain management has become one of the critical issues in today's business. It is as important as assembly line management. With global pipelines today and complex supply chain networks, it is difficult to achieve full supply chain visibility. Without complete visibility or transparency in supply chains, issues related to compliance are difficult to handle. Thus, supply chain visibility is a major challenge in the manufacturing industry.

v) **Product development and innovation.** Today's customers demand highly innovative products from manufacturers. The manufacturing industry has changed its phase from a push production system to a customer-oriented pull approach. With the level of globalisation and openness to foreign market players today, new product development and innovation has become more complex than ever. Hence, the manufacturers are required to continuously innovate in order to stay in competition with other market players.

vi) **Increasing healthcare costs.** With growing awareness of sustainability issues employee safety has become more than a regulatory requirement. Most manufacturing companies spend additional capital on system alerts and precautionary safety measures more than ever. Thus, the operating budgets of the manufacturing firms today address seriously, and it remains a critical investment for business owners.

vii) **Volatile markets.** Unpredictable market conditions today pose serious problems to the manufacturers in terms of satisfying customer needs. The shorter life-cycles of today's products make it even worse for the manufacturers, as they are faced with a challenge of bringing the right products to the market as fast as possible. Though the manufacturers intend to satisfy the customer requirements, market volatility poses a major challenge against it.

viii) **Huge data to analyse manually.** In a process-driven industry like manufacturing, a huge volume of data transfer is involved for effective reporting and decision-making. Traditionally periodic manual audits are conducted, and statistical reports are manually generated and analysed for decision-making. However, it is very difficult to handle huge data manually, and also it has the risk of human errors. Hence, the manual analysis of data is considered as a serious challenge.

Big Data and analytics can be the key to addressing the above challenges and it has the potential to have a definitive impact on manufacturing. Before discussing the impacts of Big Data on manufacturing, let's discuss what Big Data is and what its major characteristics are in the next section.

3.2.3 What Is Big Data?

The term Big Data has been in use and popularised by John Mashey since the 1990s (Lohr, 2013). Big Data includes a huge number of datasets that generally cannot be captured, curated, managed or processed within a tolerable time frame with commonly used software tools (Snijders et al., 2012). However, as the size of Big Data keeps on increasing over the years, techniques and technologies with which to integrate data and to reveal insights from complex and diverse datasets are required (Hashem et al., 2015). A standardised definition of Big Data is not available and hence we present some of its notable definitions. In 2011, the McKinsey Global Institute pointed out that 'Big Data refers to datasets whose size is beyond the ability of typical database software tools to capture, store, manage and analyze'. Gantz and Reinsel (2011) described Big Data as 'a new generation of technologies and architectures, designed to economically extract value from very large volumes of a wide variety of data, by enabling high velocity capture, discovery and/or analysis'. Bayer and Laney (2012) defined the concept, insisting that regular technologies and tools cannot store or process Big Data as they 'are high volume, high velocity, and/or high variety information assets that require new forms of processing to enable enhanced decision making, insight discovery and process optimization'. Big Data usually is encompassed by three types of data: unstructured, semi-structured and structured data, among which the main focus is unstructured data (Dedić & Stanier, 2016). Figure 3.1 presents the word cloud formed with the key terms appearing in the abstracts of Big Data-related journal articles.

Big Data is generally structured multi-dimensionally and can be characterised by the concept of various 'Vs' which grow exponentially (Shao et al., 2014). IBM scientists have identified four key aspects of Big Data as Volume, Velocity, Variety and Veracity (Taylor-Sakyi, 2016). Figure 3.2 presents the I.N.M.'s 4Vs of Big Data and the descriptions are detailed below:

i) **Volume.** This represents the quantity of data. The quantity is huge and growing every day. IBM predicts that there will be around 35 zettabytes stored by 2020. Companies may store enormous amounts of data which might be a mix of data from customers from web logs stored in databases and also the real-time information acquired through sensors in terms of production information, inventory and shipments (Chen & Zhang, 2014).

FIGURE 3.1   Cloud visualisation of key words appearing in abstracts of Big Data-related articles (adapted from http://www.professorwidom.org/bigdata, accessed online on: 20 March, 2019).

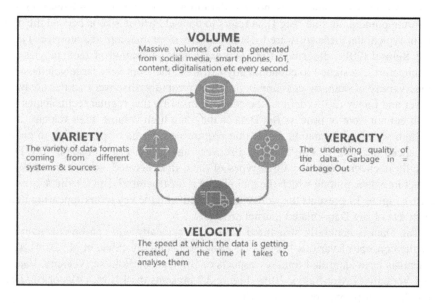

FIGURE 3.2   4Vs of Big Data (adapted from http://www.vijayraghunathan.com/big-data analytics/ accessed online on: 20 March 2019).

ii) **Velocity.** This represents the speed of data generation and transmission. The data can be processed in batch, real-time or even streamed (Assunção et al., 2015). In today's world, with inventions and increased usage of computers and smartphones, data is generated and transmitted at a faster rate. In manufacturing industries data velocity i.e., the frequency of data occurrence is to be noted carefully. For example, in production facility control cases various log data, sensor information and in–out plant environment information need to be collected and processed in a timely manner.

iii) **Variety.** This refers to the different sources from which data is acquired and the range of data types involved. The fact that data can be generated from different sources means internal and external source may be in different formats. In this, some data are structured and some are unstructured, whereas some are semi-structured (Tan et al., 2015).

iv) **Veracity.** The structured and unstructured data are generally coming from different sources and so their quality varies. Thus, the reliability and accuracy of the data are difficult to control and there is a question of relevance and credibility. Thus, veracity refers to checking the quality of data.

Other than these four major 'Vs' there are also other characteristics of Big Data identified by various researchers. Pence (2014) stressed the need for analysing an important characteristic called complexity. Complexity refers to the degree of interdependencies between the structures in the data acquired. This allows studying the small or big changes in a few components that can lead to drastic behavioural system-level changes (Satyanarayana, 2015). Visualisation is another characteristic which refers to the effective and meaningful representation of complex and large data to the users for easier understanding. Graphs, maps and figures are some visual representations which can be easily interpreted by the human brain and hence make the decision-making easier (Tang et al., 2016). The accuracy and correctness of the data which can be readily used are considered another characteristic of Big Data. This characteristic is called 'Validity' and it refers to how relevant is the data is to the organisation's current strategy to provide any useful insights. Volatility is another characteristic of Big Data and it deals with shelf-life of the data. It relates to the extent to which data can be stored and used. This is because we will have to handle a huge quantity of real-time data which are sensitive and are bound to frequent changes. The final characteristic identified is the Value of Big Data. It is already established that Big Data is a going to be a mix of structured, unstructured and semi-structured data out of which the case-specific data needs to be extracted. The corresponding data which relates to what we require is the measure of usefulness i.e., the value. The value of Big Data can be mathematically expressed as a combination of all the other characteristics as below:

$$\text{Volume} \times \text{Variety} \times \text{Velocity} \times \text{Veracity} \times \text{Complexity}$$
$$\times \text{Visualization} \times \text{Validity} \times \text{Volatility} = \text{Value}$$

3.2.4 IMPACT OF BIG DATA IN MANUFACTURING

Big Data analytics holds importance in every industry because of its ability to provide insights for decision-making in every facet of the business (Davis et al., 2012). Future manufacturing is essentially connected with Big Data. With the fourth industrial revolution reaching greater heights, manufacturing can be based on the data from I.o.T. elements and analytics along with the customer (Big) data leading to predictive approach for the future (Lee et al., 2013). Improvements in innovative processes, impact on the environment and operational efficiency are some potential benefits of intelligent manufacturing (O'Donovan et al., 2015). It is well known that advances in information technology have played a major role in the fourth industrial revolution. The rise of Big Data and software analytics had a huge influence on the contemporary industries and arguably helped in reshaping the manufacturing sector (Babiceanu & Seker, 2016). Big Data entails having a table of information about the manufacturing company in terms of product quality, total number of products in production, raw materials availability, total number of finished products, client information and orders, purchase details, and details of the manufacturing equipment, etc. (Pääkkönen & Pakkala, 2015). Companies have begun to realise that disruptive technologies like Big Data and the internet of things can provide a competitive advantage and effectively meet the business needs globally (Tao et al., 2018). Therefore, manufacturers use the industrial internet of things and Big Data to capture and make use of external and internal data to gain the competitive edge by using the insights in decision-making right from the product development stage to financial and supply chain decisions. Figure 3.3 illustrates the various data generated in manufacturing and their corresponding usage.

Let's have a closer look at how Big Data has had a positive impact on the manufacturing sector in the following subsections.

i) **Faster Cost estimation.** It is always necessary for the production team to have the cost estimation available for sales and product teams. With shorter life-cycles today and more difficult market dynamics, accurate and quicker cost estimation is needed. With I.o.T.s and Big Data analytics the data is retrieved in real-time and processed at greater speeds so that necessary information is available to make the cost estimations (Lee et al., 2013).

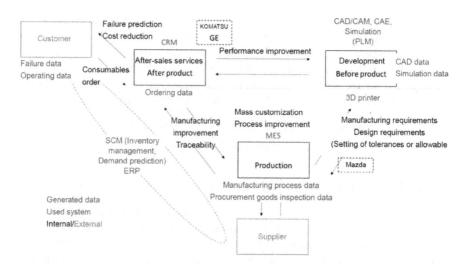

FIGURE 3.3　Generation of data and its usage in manufacturing.

ii) **Quality production.** Big Data enables manufacturers to implement Lean and Six-Sigma programs which eventually reduce wastes and variability in production. By using advanced Big Data analytics key performance indicators can be analysed and the necessary corrections can be made (Kang et al., 2016). Since most of the data can be easily collected, product quality can be improved significantly, and the quality costs can be reduced drastically. Though every single product may require a huge number of quality tests, with Big Data analytics, pattern recognition and predictive analysis can lessen the number of tests. Production line quality can be improved significantly with the help of sensor data analysis detecting manufacturing defects (Zhang et al., 2017). Through early detection, time and costs related to adjusting the production processes can be saved.

iii) **Custom product design and production.** Traditionally, manufacturers focused on production at scale for cost efficiency. Even the introduction of flexible manufacturing systems did not cover mass customisation (Zhong et al. 2017). But with Big Data and the I.o.T.s, the forecast for customised product demand can be achieved precisely. Manufacturers can gain more lead time and have the opportunity to produce customised products efficiently, similar to mass and batch production. Big Data analytics enable product engineers to access real-time data on customer preferences so that even design changes can be done at any time to meet order requirements (Wang et al., 2016).

iv) **Better demand forecast.** Using Big Data, manufacturers can anticipate the product demand accurately and more quickly, which prevents companies from wasting time and resources on inventory (Zikopoulos & Eaton, 2011). The better forecasting capabilities offer the stakeholders to step towards future business arrangements. The ability to generate better forecast reports will lead to fewer stock outs, less idle inventory and, importantly, satisfy more customers (Diebold, 2003). Manufacturers and retailers can also use social media tools to measure customer feedback and sentiments for their products and plan the product specifications accordingly.

v) **Enhanced Supply Chain Risk Management.** In today's global supply chains, the interruptions or any possible disruptions are costly and affects the manufacturer's relationships with retailers and customers (Chae, 2015). Big Data provides an opportunity for manufacturers to minimise the risks in materials delivery for production and proactively develop contingency plans for any such disruptions. Further, utilisation of Big Data analytics assists in assuring the supplier quality by assessing the performance data needed for sourcing decisions. I.o.T.-enabled systems help in efficient monitoring of inventory, location tracking and reporting of products throughout the supply chain (Zhong et al., 2015). Big Data can also deliver critical inputs into product lifecycle management (P.L.M.) and enterprise resource planning (E.R.P.) systems of the organisations (Zhong et al., 2016).

vi) **Safe work environment.** Analysing data about short- and long-term absenteeism of employees, injury, illness rates and any accident data are key performance indicators for the health and safety of the work environment. Manufacturing companies now use sensors to ensure the safety of products and labour as a part of preventive maintenance by ensuring conformity with safety regulatory requirements (Lee et al., 2013). Moreover, Big Data is the heart of Industry 4.0 and smart factories. Hence, the human workforce is replaced with robots to work under difficult and hazardous working conditions. Any disruptive changes can be sensed by, and predictive and proactive safety measures are made possible with, Big Data analytics (Loebbecke & Picot, 2015).

vii) **Proactive, predictive and preventive maintenance.** The internet of things and sophisticated sensors enable manufacturers to collect and analyse real-time data from all machinery and even from customer products (Le et al. 2013). This helps the manufacturers to act proactively by reading the patterns of data received. With operational data, predictive maintenance can be used to prevent downtime, cost-related maintenance and prolonging the lifespan of machines by preventing any permanent damage (Kang et al., 2016). Customer products nowadays are tracked and any changes in behaviour of the device are notified to the customers predicting the optimal maintenance, thereby creating better user experience and reducing maintenance and warranty costs.

viii) **Faster service and support to customers.** Big Data is mainly oriented towards analysis and prediction of acquired data. Manufacturers are looking at more complex products needing an operating system to manage the sensors onboard. Analysing real-time data from those sensors can help in customer service (Opresnik & Taisch, 2015). Even before receiving any complaints from customers, Big Data can help with framing strategies to deal with any predicted defects. Therefore, when some consumer files a complaint the company can address the issue by immediately offering action steps to deal with it.

3.2.5 How to Adopt Big Data Analytics?

Though the benefits of Big Data are clear and sound, the biggest challenge on the road to Big Data analytics is the lack of understanding of how to harness them. Companies must pay attention and commit themselves to their initiatives and use of Big Data. The effective utilisation of Big Data starts with support and commitment from top-level management in building or acquiring a sophisticated system for managing data. The companies should develop better understanding of the concepts of Big Data so that they can identify and analyse different types of data efficiently and effectively (Loebbecke & Picot, 2015). Assessing the frameworks for data generation, acquisition and management is necessary along with developing machine-learning capabilities. Usually, working with a well-established partner who can help in optimising the use of data is a better option. Generally, Big Data analytics should have a clear-cut purpose which is grounded on the right foundation and can be always conducted to draw insights for decision-making actions. Figure 3.4 exhibits the idea of effective Big Data analytics.

Big Data analytics lie on the foundation of rigid organisational governance and sophisticated technological infrastructure (Taylor-Sakyi, 2016). The companies should be committed and oriented towards Big Data with dedicated teams

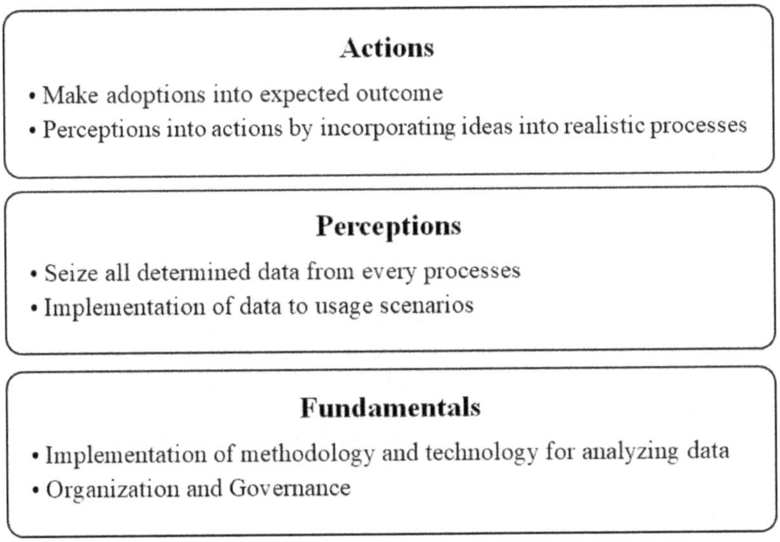

> ### Actions
> - Make adoptions into expected outcome
> - Perceptions into actions by incorporating ideas into realistic processes

> ### Perceptions
> - Seize all determined data from every processes
> - Implementation of data to usage scenarios

> ### Fundamentals
> - Implementation of methodology and technology for analyzing data
> - Organization and Governance

FIGURE 3.4 Big Data Analytics.

with skilled personnel with the knowledge and experience to work with Big Data. Technological requirements like analytical tools for statistical analysis, forecasting, regression analysis, database querying, data-storing warehouses, machine learning and data mining are required along with the physical infrastructure to run them. On this strong foundation lie the data acquisition tools to capture, store and manage data. The captured data is analysed and the foundations to provide critical deliverable insights helpful in accurate decision-making are achieved.

3.2.6 CONCLUSION

In this chapter we have attempted to summarise Big Data's role in manufacturing industries by presenting the challenges faced by today's manufacturers and how Big Data can be a solution for them. First, we summarised the manufacturing challenges present today and explained the need for Big Data. Second, a detailed description of the concept of Big Data is presented by explaining its characteristics with help of the existing literature. Third, the various advantages Big Data brings into manufacturing is detailed and finally a brief note on Big Data analytics adoption is provided. This chapter contributes to the literature on Big Data and its importance in manufacturing.

REFERENCES

A Day in Big Data. 2014. "BIG DATA for smarter customer experiences." Accessed March 14, 2019. http://adayinbigdata.com.
Assunção, Marcos D., Rodrigo N. Calheiros, et al. "Big Data computing and clouds: Trends and future directions." *Journal of Parallel and Distributed Computing* 79 (2015): 3–15.
Babiceanu, Radu F., and Remzi Seker. "Big Data and virtualization for manufacturing cyber-physical systems: A survey of the current status and future outlook." *Computers in Industry* 81 (2016): 128–137.

Bayer, M., and D. Laney. 2012. "The importance of 'big data': A definition." Gartner. Accessed June 21, 2012. https://www.gartner.com/doc/2057415/importance-big-data-definition.

Chae, Bongsug Kevin. "Insights from hashtag# supplychain and Twitter analytics: Considering Twitter and Twitter data for supply chain practice and research." *International Journal of Production Economics* 165 (2015): 247–259.

Chen, C.L. Philip, and Chun-Yang Zhang. "Data-intensive applications, challenges, techniques and technologies: A survey on Big Data." *Information Sciences* 275 (2014): 314–347.

Davis, Jim, Thomas Edgar, et al. "Smart manufacturing, manufacturing intelligence and demand-dynamic performance." *Computers & Chemical Engineering* 47 (2012): 145–156.

Dedić, Nedim, and Clare Stanier. "Towards differentiating business intelligence, Big Data, data analytics and knowledge discovery." In *International Conference on Enterprise Resource Planning Systems*, Springer, Cham (2016): 114–122.

Diebold, Francis X. "Big Data dynamic factor models for macroeconomic measurement and forecasting." In *Advances in Economics and Econometrics: Theory and Applications, Eighth World Congress of the Econometric Society*, edited by M. Dewatripont, L.P. Hansen, and S. Turnovsky (pp. 115–122). Cambridge University Press, Cambridge, UK, 2003.

Durak, Umut. "Flight 4.0: The changing technology landscape of aeronautics." In *Advances in Aeronautical Informatics*, Springer, Cham (2018): 3–13.

Gantz, John, and David Reinsel. "Extracting value from chaos." *IDC Iview* 1142, no. 2011 (2011): 1–12.

Hashem, Ibrahim Abaker Targio, Ibrar Yaqoob, et al. "The rise of 'Big Data' on cloud computing: Review and open research issues." *Information Systems* 47 (2015): 98-115.

IBM. "What is Big Data?" 2014. http://www- 01.ibm.com/software/data/bigdata/what-is-bi gdata.html.

Kang, Hyoung Seok, Ju Yeon Lee, et al. "Smart manufacturing: Past research, present findings, and future directions." *International Journal of Precision Engineering and Manufacturing-Green Technology* 3, no. 1 (2016): 111–128.

Lee, Jay, Edzel Lapira, et al. "Recent advances and trends in predictive manufacturing systems in Big Data environment." *Manufacturing Letters* 1, no. 1 (2013): 38–41.

Loebbecke, Claudia, and Arnold Picot. "Reflections on societal and business model transformation arising from digitization and Big Data analytics: A research agenda." *The Journal of Strategic Information Systems* 24, no. 3 (2015): 149–157.

Lohr, Steve. "The origins of 'Big Data': An etymological detective story." *New York Times* 1 2013.

McNeil, Patrick. *Web Designer's Idea Book, Volume 4: Inspiration from the Best Web Design Trends, Themes and Styles.* Simon and Schuster, Cincinnati, OH, 2014.

O'donovan, Peter, Kevin Leahy, et al. "Big Data in manufacturing: A systematic mapping study." *Journal of Big Data* 2, no. 1 (2015): 20.

Opresnik, David, and Marco Taisch. "The value of Big Data in servitization." *International Journal of Production Economics* 165 (2015): 174–184.

Pääkkönen, Pekka, and Daniel Pakkala. "Reference architecture and classification of technologies, products and services for Big Data systems." *Big Data Research* 2, no. 4 (2015): 166–186.

Pence, Harry E. "What is Big Data and why is it important?" *Journal of Educational Technology Systems* 43, no. 2 (2014): 159–171.

Satyanarayana, Lenka Venkata. "A survey on challenges and advantages in Big Data." *IJCST* 6, no. 2 (2015): 115–119.

Schmidt, Eric, and Jared Cohen. *The New Digital Age: Reshaping the Future of People, Nations and Business.* Hachette, United Kingdom, 2013.

Shao, Guodong, Seung-Jun Shin, et al. "Data analytics using simulation for smart manufacturing." In *Proceedings of the Winter Simulation Conference 2014*, IEEE, Savannah, GA (2014): 2192–2203.

Snijders, Chris, Uwe Matzat, et al. "'big Data': Big gaps of knowledge in the field of internet science." *International Journal of Internet Science* 7, no. 1 (2012): 1–5.

Tan, Kim Hua, YuanZhu Zhan, et al. "Harvesting Big Data to enhance supply chain innovation capabilities: An analytic infrastructure based on deduction graph." *International Journal of Production Economics* 165 (2015): 223–223.

Tang, Jian, Jingzhou et al. "Visualizing large-scale and high-dimensional data." In *Proceedings of the 25th international conference on World Wide Web, International World Wide Web Conferences Steering Committee* Montréal, Québec, Canada (2016): 287–297.

Tao, Fei, Jiangfeng Cheng, et al. "Digital twin-driven product design, manufacturing and service with Big Data." *The International Journal of Advanced Manufacturing Technology* 94, no. 9–12 (2018): 3563–3576.

Taylor-Sakyi, Kevin. "Big Data: Understanding Big Data." *arXiv preprint arXiv:1601.04602* (2016).

Walport, M. "The Internet of Things: Making the most of the second digital revolution, a report by the UK government chief scientific adviser." *Technical Report*, 2014. https ://www.gov.uk/government/uploads/system/uploads/attachment_data/file/409774/14-1 230-internet-of-things-review.pdf.

Wang, Shiyong, Jiafu Wan, et al. "Towards smart factory for industry 4.0: A self-organized multi-agent system with big data based feedback and coordination." *Computer Networks* 101 (2016): 158–168.

Zhang, Yingfeng, Shan Ren, et al. "A Big Data analytics architecture for cleaner manufacturing and maintenance processes of complex products." *Journal of Cleaner Production* 142 (2017): 626–641.

Zhong, Ray Y., George Q. Huang, et al. "A Big Data approach for logistics trajectory discovery from RFID-enabled production data." *International Journal of Production Economics* 165 (2015): 260–272.

Zhong, Ray Y., Stephen T. Newman, et al. "Big Data for supply chain management in the service and manufacturing sectors: Challenges, opportunities, and future perspectives." *Computers & Industrial Engineering* 101 (2016): 572–591.

Zikopoulos, Paul, and Chris Eaton. *Understanding Big Data: Analytics for Enterprise Class Hadoop and Streaming Data.* McGraw-Hill Osborne Media, USA, 2011.

3.3   NETWORKING FOR INDUSTRY 4.0

Lokesh Singh, Someh Kumar Dewangan, Ashish Das, and K. Jayakrishna

3.3.1   INTRODUCTION TO NETWORKING FOR INDUSTRY 4.0

The pattern is that an ever-increasing number of advances are made to help us in ordinary undertakings. Be that as it may, for this to occur, we should utilise them proficiently and effectively [8]. These smart devices enable the coordination of various strategies; for instance, communication devices themselves, production data, natural data, equipment status and various different organisations. Industry 4.0 is turning into a reality for some and is carrying with it a large group of advantages for shoppers and manufacturers, including lower costs, quicker production, better asset effectiveness, higher quality control and more significant detectability. Guaranteeing adequate execution is a key prerequisite for Industry 4.0 achievements, so manufacturers intending to move towards these smart factory ideas need to consider the effect that their selected network technology may have. Unmistakably, mechanical control networks are fundamental to Industry 4.0 and will be founded on Ethernet, since it is an international standard that guarantees interoperability of modern and commercial networks and makes the industrial internet of things (I.I.o.T.) conceivable. In any case, regardless of the intermingling of principles which has expanded the interoperability of networks and hardware from multiple vendors, creating hardware which can work on all accessible mechanical network benchmarks is as yet illogical for producers of modern products [9]. In Industry 4.0, there may be numerous brilliant devices, devices that require organised associations to connect with each other [15]. We have defined some requirements for the implementation of Industry 4.0.

3.3.1.1   Mass Communication

This communicational model is never dependent on 'mass', but is one of a 'network'. Notwithstanding the structural change and the powers that shape it [20], one can likewise distinguish a lot of relevant changes that are, on occasion, the consequence of the appointment of this new networked communication programme and, on different occasions, an indication of the advancement of another media framework with another character.

3.3.1.2   Flexibility

The manufacturing framework is influenced by a wide scope of factors. In this, we have six run-of-the-mill manufacturing frameworks; there are many other manufacturing frameworks characterised by engineers, for example, P.C. incorporated manufacturing framework, reconfigurable manufacturing framework, etc. [13].

In contemporary manufacturing undertakings, regardless of whether they are adaptable or conventional, a combination of the individual utilitarian territories of a venture into an integrated P.C.-supported framework is by all accounts vital. Such coordination empowers through automation recognition of practically the majority of the organisation's capacities by P.C.s or devices working under their watch. This removes human participation in the implementation of innovative and helper tasks, for example, manufacturing, transport, capacity, control of numerically controlled devices and implementing information frameworks to control and observe procedures [12, 2].

3.3.1.3   Factory Visibility

Current frameworks are moving towards issues of asset practicality, reduced time to exhibit, formation of adjusted things and the interconnecting of system infrastructural components [9]. The objective of every cutting-edge framework and every mechanical upset is to create productivity.

3.3.1.4   Connected Supply Chain

The supply chain today is a movement of, all things considered, discrete, siloed steps taken through advertising, product improvement, manufacturing and distribution, finally under the control of the customer [35]. This system will depend on different key innovations: incorporated orchestration and execution systems, logistics, detectable quality, independent coordination, smart acquisition and warehousing. The result will enable organisations to react to disruptions in the supply chain and even imagine them by showing the complete network [57].

3.3.1.5 Energy Management

Energy management is the total of the considerable number of measures that are arranged and implemented so as to guarantee a base energy use for the required performance [78]. What's more, in numerous production plants in the manufacturing industry there is a reasonable potential for saving energy: for instance, tweaked automation arrangements can achieve the improvement of systems and processes. Therefore, plants are used more productively and, consequently, unnecessary energy utilisation because of downtime is avoided [61].

Chosen points of interest of productive energy management:

- Transparency of energy streams and energy use
- Energy cost decreases
- Optimised energy acquirement
- Risk minimisation against fluctuating energy costs
- Environmental insurance and feasible management.

In any case, the reason for increasing energy proficiency is the systematic collection and analysis of energy use: in order to reveal potential reserve funds and indentify changes, the whole energy stream of the organisation must be recorded and documented [31]. Inside the structure of modern industry considering the current scenario, for instance, a cross-merchant arranged creation of almost all information required for this is starting to be available now and should not be accumulated in an expensive way [35].

3.3.1.6 Creating Values

An esteem network is a lot of connections among associations as well as with people cooperating with one another to enable the whole group to profit [19]. An esteem network enables members to purchase and sell products just as they share information. These networks can be visualised with a straightforward mapping device showing members and connectors' relationships [91].

3.3.1.7 Remote Monitoring

The new plan of activity is apparently rapidly dying down into an organisation industry instead of an item or generation industry [89]. This Industry 4.0 and the mechanised networking of manufacturing strategies will make it possible to fabricate items at lower cost, and in a way that is progressively versatile, vitality capable, asset-saving and customisable [16].

For machine manufacturers and sellers, remote internet access will engage them to all the practically certain total the assistance plan sold with their things to pass on ruling client association and a greater proportion of contraption accessibility or up time. Furthermore, obviously, keeping up a crucial separation from personal time is a focal goal for any assembling venture [65].

3.3.1.8 Proactive Industry Maintains

Industry 4.0 is driving change associated with manufacturing, conveying with it all the changes, opportunities and difficulties that represent [43]. It interfaces the inventory network and the E.R.P. framework explicitly to the generation line to outline a joined, robotised and, conceivably, self-deciding manufacturing model that improves usage of capital, raw materials and human resources [49]. Industry 4.0 furthermore presents a totally unique course for manufacturers to improve businesses and change into administration providers using data [56]. Like making various items on a comparable mechanical creation framework, in and of itself, this isn't new. Organisations routinely change inside information into administration. However, it extends this possibility to items like electric motors that control cranes that, beforehand, were never seen as endpoints on a system that could create saleable data [4].

3.3.1.9 External Communication for Devices through Gateway S.D.N.

System-based communication is a vital segment for changing the set-up of the present-day creation pyramid into an organised system of appropriated systems. It will empower the patching up of current business relations and help to make new systems including networks [14]. To structure the diverse situations and necessities for system-based communication, a central field of activity of the network is to draw up a related reference appearing as an upgrade to R.A.M.I. 4.0 [3]. An important part of this reference is contained in administrations through which applications can mastermind the required Industry 4.0-satisfying system parameters [7]. Applications should more likely than not orchestrate their complete communications using methods for Industry 4.0-satisfying interfaces. The system assets sent can in like manner be facilitated by methods for non-Industry 4.0-satisfying interfaces. Diverse Wi-Fi and Ethernet-based mechanical communication systems have been essentially used in this division as of not long ago, particularly for applications with incredibly high persistent quality and idleness necessities. The Ethernet standard is directly being refreshed for guaranteed process terms under IEEE 802.1 in the TSN Group. Wireless networking courses of action are upgraded with different general benchmarks for remote neighbourhoods, for instance, Wi-Fi, DECT ULE, Bluetooth or 6LoWPAN, which have been not entirely unequivocally balanced (reinforced) for mechanical use (for instance, various transmissions for improved reliability) [8]. There are similarly high standards for adjacent remote networks underway conditions, for instance, WirelessHART, according to IEC 62591. Up to this point, in any case, the open remote rules have as a general rule failed to meet the stringent idleness and reliability requirements for application in a generation cell, for instance for development control [58]. These are being upgraded under IEC 62948, for example (Industrial networks: wireless communication system and communication profiles) [34].

3.3.1.10 Connection and Management of Data in the Cloud

New worldwide necessities, for instance, environmental concerns, similarly benefit from the use of these movements, which have changed the strategy for manufacturing in business with the use of green items and sustainability. Based on these movements and improvements, the standard explanation behind our chapter is to propose a plan of communication that can help in the industrial progress of enterprises, likewise called Industry 4.0, making the communication progressively agile and efficient [89]. We are as of now the key partner of a large number of organisations with regard to exhaustive investigation solutions. Consolidating our application know-how and our investigation ability we offer platform-free solutions, equipped to the particular needs of our customers [56]. We generate included value, for instance, by getting conclusions from coordinating information investigations in the context of solid applications. Based on machine learning conduct, we can perceive potential faults all the while and keep them from happening. Thusly, unplanned downtime can be fundamentally diminished, bringing about an expansion in machine productivity [90]. Digitalisation opens up totally new conceivable outcomes and opens doors for sorting out, checking and controlling manufacturing methods. At the same time [67], Industry 4.0 is thinking of totally new prerequisites that must be tended to. The engineering industry, for example [5], has perceived that digitalisation solutions are playing an pivotal job in an organisation's success [77]. This affects the development of individual administrations as much as information-based plans of action or prescient upkeep systems. It is our expressed plan to free you of these uncertainties and, filling in as partners, go with you on your individual journey towards the smart factory [84]. To supplement the amalgamation of automation and digitalisation, the distinctive part of our suggestion is the special blend of hardware, programming, engineering and consultancy. Utilise the opportunities of the current mechanical boom and work with us to develop your solutions for the future [54].

3.3.1.11   Dynamic Management of Smart Devices

So as to meet above chose preconditions of association engineered mastermind foundation, normal framework topology configuration would be established on star or ring topology [8]. Mechanical highlights of such framework are versatility, receptiveness, execution (most extreme and speed), reasonableness, ampleness and security. S.N.M.P. is kept up by most business N.M.S.s and many systems administration gadgets, including switches, switches, servers and workstations. S.N.M.P. has extended wide clarity in view of its simplicity and considering that it is certainly not difficult to finish, present and use. The R.M.O.N. ready store up gives a framework official a chance to set configuration boundaries for framework parameters and arrange features to in this manner give alerts to N.M.S.s [43]. R.M.O.N. also bolsters getting bundles (with channels at whatever point required) and sending the bundles to an N.M.S. for a demonstration examination [52]. To guarantee high framework accessibility, the administrators' mechanical bundles ought to strengthen various highlights which can be utilised for execution, issuing, planning, securing and accounting for the framework. In a perfect world, the framework should in a like way join information to perceive structures that can predict a potential failure with the target that a framework boss/maker can make a move to address before a fault happens. Checking gadgets are from time to time reliant on adhering to displays and gauges: accept that the most authentic method should be the method which puts the focus on the result. For straightforwardness, allow us to portray the unpredictable variable X as equal to the amount of the clock cycle related to the execution of the method m. Each acknowledgment x of this self-assertive variable, is proportionate to

$$x = Pki = 1(ci) + Pki = 1(ni),$$

with k to such an extent that $Pk\ i =1(di) = m$, with i being the ith succession of futile/valuable guidelines and with di and ni the ith realisations separately of the irregular factors D and N. We have

$$x = m + PkI = 1(ni).$$

We think that, since $m \gg D$, x could be approximated by $x \sim m + Pq\ I = 1(ni)$ with $q = 2 \bullet m/ (D + 1)$, in these conditions, the depth of likelihood of X pursues an ordinary dispersion $(\mu X, \sigma X)$ with:

$$\mu X = m + q \cdot \mu N = m \cdot (1 + N/2)(2)\sigma 2X$$
$$= q \cdot \sigma 2N = m \cdot N \cdot (N + 2)6 \cdot (D+1)$$

(3.3)

For simplicity, we consider, as proposed in [13], that m is consistently appropriated (with the likelihood 1 out of 2 · σX) between $m - \sigma X$ and $m + \sigma X$. F.S.C.A.: in these conditions, the S.C.A. peak is diminished by a factor of 2 · σX and the quantity of bends important to recover the key increments by a factor $4 \cdot \sigma\ 2\ X$. Yet, by utilising the sliding window strategy (which comprises recreating the peak by incorporating the utilisation bends on 2 · σX tests) additionally portrayed in [13], this sparing as far as the number of power bends is just equivalent to 2 · σX. F.D.F.A.: in a request to realise a D.F.A., we guess that the assailant can target clock cycles containing between $m - \sigma X$ and $m + \sigma X$. In these conditions, he has one possibility out of 2 · σX to adjust the guidance m. The quantity of broken realisations is subsequently expanded by a factor of 2 · σX. FTime: Equation (3.3) shows that the calculation time is expanded by a factor $(1 + N/2)$. F.N.R.J.: because we think that both helpful and false directions consume similar energy, the utilisation of the circuit is additionally expanded by a factor of $(1 + N/2)$[52].

F SCAIDI = (if $N = 0\ 2 \cdot q\ m \cdot N \cdot (N+2)\ 6 \cdot (D+1)$
AIDI = F SCAIDI F
TimeIDI = $1 + N/2$ F
NRJIDI = F T imeIDI

In these conditions, the hard and fast energy usage of the circuit is extended by a factor [5]

$$(1 + \alpha \cdot R). \text{ F SCARP G} = 1 + R \text{ 2 F DF}$$
$$\text{ARP G} = 1 \text{ F T}$$
$$\text{imeRP G} = 1 \text{ F NRJRP}$$
$$G = (1 + \alpha \cdot R)$$

At the point when the unmistakable are joined, factors adding various countermeasures are essentially copied for other portion-channel attacks, for the period of estimation and the energy utilisation [76]. It is equal to the plausibility procured in the essential occasion (for instance $1/(2q \cdot 2 \cdot \sigma X)$) up to the overabundance measurement short one [32]. It shows the estimations of these segments.

3.3.1.12   Feed of Data and Automatic Decision-Making

Information, especially when amassed, can uncover a lot about an individual [4]. For instance, when somebody calls their closest partner, for example, there is information about valuable attributes, leads, regions and contacts; likewise with logically promoted information on training, for example, A.I. When made, profiles can lay out the interpretation behind central power. In particular, smart planning and information structures are required that connect with regular managing and can supervise reliable updates. Additionally, methods ought not rely on a priori suppositions with respect to information load, landing rate or purposes of restriction, since leaked information may be untrue [25]. Additionally, we recognise that streams can't be taken care of completely in light of their tremendous size and that in this way, on-the-fly managing and acumen is required [16]. For specific frameworks, there is no unequivocal portrayal of how they carry on in the introductory reorganisation, for example, right when the stream begins and the important information comes in. While a few calculations undoubtedly won't require information from the past, others genuinely rely on it and will require it to have some place in the scope of an opportunity in order to wind up stable and meaningful [23].

3.3.1.13   Optimisation of Customers Directly with Industry 4.0

Industry 4.0, intelligent manufacturing and the mechanical I.I.o.T. are current examples that are fundamental to manufacturing advancement and productivity [54]. Industry 4.0 is the fourth modern revolution and is normally a picture of a system among things and people all through the manufacturing methodology. Brilliant manufacturing is the path towards constraining human participation and perhaps utilising human mental capacity when it matters [57]. Though a segment of the objectives for Industry 4.0 are extremely desirable, the hope is that it can progress into man-made mental aptitude and modernised fundamental authority, close to faultless mechanical robotization, and worthwhile human compromise, and have manufacturing workplaces that are completely interconnected and 'smart', from rough materials to finished products [78]. Streamlining of manufacturing information comes through examination, re-enactment, perception and security support, etc. In the end, the objective is to lessen costs and improve quality. The improvement and computerisation of innovation doesn't require an all-out ejection of people from the method; it's an inconceivable inverse. Unique workplaces require stimulating examination and toolsets. This creates an open pathway for mechanical inventors, information researchers, manufacturing professionals and analysts, and requires another kind of information-driven manufacturing specialist to decide on improvements and optimisation [83].

3.3.2 History of Networking in Industry

This is also true for current control systems. From the beginning, control of gathering and methodology plants was done precisely – either physically or using weight-driven controllers. As discrete equipment wrapped up standard, the improvement towards cutting-edge structures accomplished the prerequisite for new trades shows to the field moreover as between controllers [21]. It will be seemed, by all accounts, to be accessible day systems spread an extensive space and are of loosening up immensity to fields, for instance, assembling and power age [12]. They are extremely certain and make usage of a social affair of conventions that have been exclusively fitted to fulfill the careful necessities that work out as intended as a result of executing consistent control of physical gear [79]. The most pressing barrier is that mechanical systems are associated with physical machines in some buildings and they are used to control and screen genuine exercises and conditions [1] (Table 3.1).

3.3.3 Need for Networking in Industry

Major sectors of the economy are experiencing a change as robots, vehicles, and modern control systems become connected to the internet [24]. This pattern is empowering exceptional dimensions of computerisation at the plant level, yet all through

TABLE 3.1
Industrial and Conventional Network [45]

	Industrial	Conventional
Primary function	Physical equipment	It should be D.T.P. (Data processing and transfer)
Applicable domain	Control of physical equipment manufacturing, processing and utility distribution	Corporate and home environments
Hierarchy	Deep, practically isolated chains of command with numerous conventions and physical benchmarks	Shallow, integrated hierarchies of protocol with uniform convention and physical standard use
Failure severity	High	Low
Reliability required	High	Moderate
Data composition	Small packets of periodic and aperiodic traffic	Large, aperiodic packets

business processes [46]. The greatest barrier to legitimate Industry 4.0 execution is our capacity for machines and devices to cooperate and share data. The expectation is that this can be maintained from a strategic distance through institutionalisation of machine-to-machine communication, I.I.o.T. communication, and ideally, programming database and communication conventions for M.P.R., E.R.P., M.E.S. and so on. When this is accomplished, we would then be able to see manufacturing transform into a completely separate and compatible arrangement of programming and equipment. This takes into consideration the joining of the best segments in equipment and programming from around the globe to realise the best solutions [88]. This change achieves really connected offices and enterprises, and what once cost billions to set up is presently progressing towards becoming standard. The majority of this requires the makers to make gradual advances and not to see Industry 4.0 as an inaccessible assignment, but rather as a steady change in mentality, and take small steps towards digitisation, sensorisation and optimisation [56]. The protocol has undergone major evolution since its inception back in the 1970s, and as well as being the number one choice for office-based networks, it has since staked a claim to be an incredibly capable and robust protocol and infrastructure for many industrial networks, including manufacturing, processing [64], oil and gas, and the food and beverage industries.

Here enters the industrial Ethernet, bringing time-critical synchronisation, high network reliability and efficiencies suited to motion control and safety applications. Unlike in an office setting, Ethernet for industrial use requires higher degrees of toughness in the shape of connectors and cabling, but more important is the need for a much higher degree of determinism which is gained from industry-specific protocols, including among others, PROFINET, EtherNet/IP and EtherCAT [22].

The communication capabilities of industrial Ethernet go far beyond the bounds of a factory's walls or an oil rig's superstructure; this information can be shared in both directions across all levels of the value chain, enabling a head office in another country access real-time drilling information, or a maintenance engineer six time zones away to fine-tune the machine his company delivered and installed [51]. It also allows traditionally separate systems, such as industrial process, energy infrastructures and building management systems, to interact. To exploit all these benefits, it makes sense that companies should by now be developing some form of Ethernet strategy. Many of the leading automation suppliers already have significant assets in place and help programmes to guide companies wishing to adopt smart factory or connected operations. Many existing networks have grown organically and so the migration path is not always clear-cut, but it is not as difficult as you may think. By segregating your enterprise or network into primary nodes and application areas, the job becomes more straightforward as the traffic (Figure 3.5).

With a factory network, mechanical organisations can construct one normal, united and tough plant-to-business arrangement. Along these lines, you can:

- Connect manufacturing plant mechanisation and control systems to I.T. systems with norms based I.P.-networking
- Improve operational proficiency and cut expenses with exceedingly secure remote access

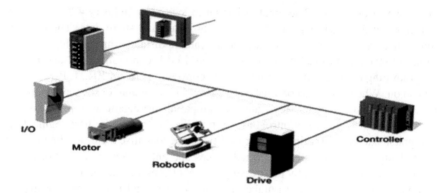

FIGURE 3.5   Factory network.

- Speed reaction times to issues on the plant floor
- Mitigate chances by improving system uptime and hardware accessibility
- Protect basic manufacturing resources with industry-driving security highlights
- These advantages will enable you to settle on progressively important business choices supported by constant data from the plant floor [19]
- Industry-driving enterprise networking arrangements conveyed crosswise over mechanical stages
- Scalable, profoundly secure, ongoing execution consistency and remote access
- Full life-cycle of administrations to enhance, organise, execute and disentangle and computerise activities
- Expertise dependent on Cisco and Rockwell Automation's Converged Plantwide Ethernet-approved engineering
- Access to the Cisco associates' ecosystem including Rockwell Automation, Honeywell and Emerson It's Connected Factory Wireless offers adaptable, plantwide communications between things, machines, databases and individuals on the plant floor. Making a protected, connected, simple-to-oversee modern wireless system cuts costs while expanding productivity and yield.

3.3.4   Vision for Networking in Industry

An autonomous system was once seen as a major aspect of an idealistic perfect network, yet one that is far away from what's to come. It would make up the foundation of all that we did, dealing with a hyper-connected world in which everything, from the particulars of knowing when the milk in the refrigerator required replacing, to the capacity of the system to naturally ramp benefits up or down, without the requirement for human intervention.

- The globally perceived pioneer of technical, interdisciplinary alumni instruction in data networking, data security and versatility.

- We will pull in the best performing, most technical, inquisitive and dedicated people to our projects and set them up for authority in their field and the larger societal setting.
- We will attract, keep, support and develop a diverse undergraduate population.
- Our graduates will be the most sought after by industry, academia and government in their individual fields.

3.3.5 INITIALISATION OF AND BASIC MATTERS ABOUT NETWORKING IN INDUSTRY

In the start of current exchanges, computerisation frameworks called fieldbus systems were made, initially with no outside help to overcome the barriers recognised [12].

The purpose of the combination of appropriated automation structures is, on an exceptionally fundamental level, the robust exchange of data. Any undertaking to administer shapes openly of tenacious human correspondence requires, in a surprisingly wide sense, the surge of data between some kind of sensors, controllers and actuators [1]. After the partner of steam limit with moderate specialists from hard physical work and the improvement of liberal scale manufacturing reliant on division of work, the presentation of motorisation progression was what is today a noteworthy piece of the time called the third industrial revolution [2]. To engage data exchange, an epic number of present-day correspondence frameworks made dependably, starting from the 1980s. It is basic that these overhauls, as a last resort, got and fit new improvements moving in various fields, overwhelmingly in the data and correspondence headway (I.C.T.) world. Ethernet, remote frameworks, or web propels are instances of this cross arrangement. These new advances made new openings for making data exchange consistently secure. Hence, robotisation systems could turn out to be persistently staggering, besides. Other than joined the physical layers, for instance, in transport systems like controller zone compose (CAN), PROFIBUS, or INTERBUS, to give a few models. Figure 3.1 reviews the timetable of this movement and engravings accomplishments in various progression fields essential for the improvement of correspondence in computerisation. Things changed around the millennium, when internet improvement twisted up undeniably understood, and I.T. ended up being comprehensively used and a dash of standard ordinary closeness. In automation, this enabled another surge of Ethernet-based frameworks that got fundamental headway from the I.T. world. The non-appearance of genuine advancing points of confinement in standard Ethernet, regardless, kept the progress of one single Ethernet answer for computerisation purposes and drove, again, to the move of submitted approaches [13], [14].

Mechanised change is the point of convergence of the fourth mechanical and 5G sifts through foundations will be key supporting assets. In the following decade, the assembling business is relied on to make toward a course relationship of creation, with related stock (things with correspondence limit), low-essentialness shapes, aggregate robots and combined assembling and collaborations. These musings are obviously exemplified under the Industry 4.0 point of view and incited two or three application conditions depicted by a working get-together of the German Plattform Industrie 4.0

[39]. One driving application condition is to lay out an arrangement of geographi-cally passed on assembling plants with versatile adjustment of age limits and sharing of focal points and assets for improve request satisfaction. Notwithstanding vari-ous things, a dependable wide region correspondence is required for this use case. Considering these changes, vertical ventures will have improved specific limits open to trigger the progression of new things and organisations. Quite a while back, a non-exclusive correspondence programme was created containing the three layers of systems, middleware and applications [42].

3.3.6 REQUIREMENT, ASSESSMENT AND METHODOLOGY OF NETWORKING IN INDUSTRY

The advancement in telecommunication innovations and the prerequisites for improved convenience and viability that is economically smart in current Industrial Control Systems (I.C.S.), for example, the Supervisory Control, appear to have encountered the evolution towards changing to the internet of things and cloud enrolling progresses. Subsequently, it is essential to observe and research the secu-rity vulnerabilities and the deficiencies of these systems to make security game plans and protection segments. Additionally, as innovation grows, so do the risks of cyber-attacks, and it's necessary to have practical disclosure and protection measures for timely reporting and initial responses to attacks. Generally, three countermeasures are open to check the S.C.A.D.A. systems, which are: 1) to recognise known secu-rity sites at the fringe of the system by using a couple of security instruments, for instance, firewalls and systems that perceive intrusions and viral attacks; 2) to dis-member the common stream of data in systems and to examine kind control work in S.C.A.D.A. .mastermind in the undertaking to recognise risks in light of change or hurting attempts, taking everything into account, 3) which is a necessary approach, to clear out the vulnerabilities in the control system structures and executions by performing specialised looking into tests, for example, infiltration tests.

A security appraisal is an ordinarily used practice that measures the present cyber-security position of a data system. Generally, the security evaluation offers proposals and guidelines to redesign its security and protection instruments to mitigate risks and keep up a vital separation from potential security threats. Generally, most inter-net-protection systems, similarly as with interconnected systems and applications present distinctive security risks and potential perils. Security specialists address these security risks through risk appraisal, weakness evaluation and infiltration test frames. On the other hand, entrance testing involves abusing the perceived weak points of the system to evaluate the security of its IT establishment. The entrance analyser covertly mimics potential activities of a dangerous aggressor by attacking the recognised vulnerabilities when assessing the adequacy of implemented security countermeasures, similarly as with confirmation instruments in the objective system.

A great deal of criteria was used to choose whether a standard could be consid-ered for review:

- The standard/rule is straightforwardly open and is written in English language

- The standard/rule is distributed by a standard body or authoritative office
- The standard/rule must be executed for/or associated with respect to I.C.S./S.C.A.D.A. systems
- The standard/rule shows a point by point delineation of the proposed security appraisal.

3.3.6.1 Methodology

In this manner, 11 models and guidelines had been picked subject to the described essentials. The recognised appropriations were examined in detail to find the most sensible weakness appraisal system that could capably manage the evaluation process, close to learning consistence with N.E.R.C. principles. The rest of this section presents a brief audit of each assessed strategy. The security evaluation is the key method in any data security program. Regardless, assessing the security of essential organisations and modern control systems is more snared and testing than general standard I.T. systems. This is a direct result of the manner in which that establishment's control systems' strategies and equipment are increasingly weak and even more vulnerable to attack. For example, using a standard tally and looking at procedures on mechanical control systems can result in discouragement with serious results in the physical condition. In this manner, assessing the security of mechanical control and essential establishment systems requires attentive orchestration and execution. This area rapidly demonstrates security appraisal challenges in mechanical systems to be considered in the midst of the organising stage for progressively extensive security evaluation. Possibly the most troublesome issue of the I.C.S. security evaluation is a result of its multi-faceted nature, generous scale and heterogeneity. Being a cyber-physical system, with the scale and multi-faceted design of I.C.S. systems and the communication innovations that they are connected with, making organising, executing and keeping an eye on cyber- and physical security appraisals becomes a very troublesome issue.

To in all likelihood perceive how Industry 4.0 transformed into the present well-known version, a look at its progenitors may give us a point of view on how this revolution is explicitly exceptional. This diagram exhibits a timetable of the advancement of manufacturing and the mechanical area overall (Figure 3.6).

3.3.6.1.1 *First Industrial Revolution*

The mechanical miracle in Britain came in to bring machines into age before the completing of the 18th century (1760–1840). This included going from the manual age to the utilisation of steam-filled engines and water as a wellspring of power. This supported growth wonderfully and the explanation 'creating plant' changed into really typical. Manufacturing benefitted greatly from such changes and was the first endeavour to put such structures in place. This fuelled a large part of the British economy at the time.

3.3.6.1.2 *Second Industrial Revolution*

The ensuing industrial revolution dates to some point in the range of 1870 and 1914 (anyway, its features can be traced back to 1850) and appeared past systems, for instance, transmits and railroads into tries. Possibly the portrait typical for that

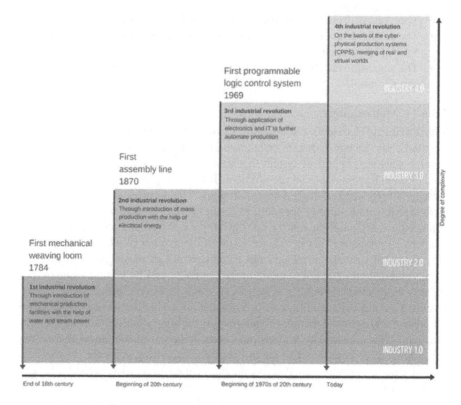

FIGURE 3.6 Network evolution.

period was the presentation of colossal scale manufacturing as a key method of the era when all is said and done. The improvement of the built hiding, moreover scratching such period as science was in a to some degree unforgiving state by then. In any case, such powerful approaches to manage and regulate industry were put to an end with the start of the First World War. Mass growth, doubtlessly, was not put to an end; regardless only types of progress inside a particular setting were made, none of which can be called modern transformation.

3.3.6.1.3   Third Industrial Revolution

Perhaps the third industrial revolution is fundamentally more normal to us than the rest, as by a wide margin most people living today have witnessed the move toward automated advances in progress. By the by, the third mechanical change is dated to some point in between of 1950 and 1970.

It is routinely referred to as the Digital Revolution, and came about as the change from fundamental and mechanical frameworks to technological ones.

Others consider it the Information Age, however. The third furious was a snappy postponed result of the enormous headway in P.C.s.

3.3.7   Advantages, Disadvantages and Limitations

3.3.7.1   Advantages of Industry 4.0

Improvement: It is a key favoured viewpoint of Industry 4.0. A smart factory containing hundreds or even a larger number of smart devices that can self-streamline age will impel an essentially zero individual time in progress.

3.3.7.2   Difficulties Confronting Industry 4.0

Security: Maybe the most problematic part of Industry 4.0 systems is the I.T. security issue. This electronic connection will offer openings to security breaches and information leaks. Cyber-attacks ought to be considered as well.

Capital: Such a change will require a tremendous enthusiasm for another development that doesn't sound shameful. The decision to roll out such improvement should be made at C.E.O. level. Additionally, still etc.

Work: While notwithstanding all that it remains before timetable to assess on business conditions with the selection of Industry 4.0 surrounding, it is protected to express that stars should confirm unique or an all-new strategy of limits. This may help business rates go up yet it will similarly evacuate an essential segment laborers.

Security: The client's stress is also the manufacturer's. In such an interconnected industry, manufacturers need to aggregate and separate information. To the client, this may look like a risk to his security. This isn't just restricted to clients. Small or large associations who haven't shared their information in the past should work their way to a dynamically immediate condition. Overcoming any issues between the purchaser and the creator will be an enormous test for the two parties.

3.3.7.3   Limitations

In any case, there are so far various difficulties that should be dealt with deliberately to ensure a smooth development. This should be the purpose of combination of wide undertakings and governments the same. Pushing assessment and experimentation in such fields is essential.

3.3.8   Conclusion and Future Scope

3.3.8.1   Conclusion

In the different level setting, security if all else fails suggests budgetary undertakings with no section of hypothesis (R.O.I.), which clears up the manner by which that security scenes are regularly overseen once the improvement system is done and when a principal security event happens, with veritable and energetic impacts to the association. In any case, this framework is remarkable what's all the more dependably fails to pass on a reliable response for the significant issue. In addition, it might condition the wellspring of separation among contenders and ruin the high grounds and the dynamic trust in their business works out. Moving towards Industry 4.0 brings accumulated technical difficulties, with high impacts in various areas in the present manufacturing industry, particularly in the security area.

3.3.8.2 Future Scope

Industry 4.0 has a lot to guarantee concerning pay rates, speculation and innovative sorts of progress, regardless business still remains a legend among the most anomalous bits of the new current shock. It's broadly more difficult to assess or check the potential business rates.

Industry 4.0 might be the pinnacle of mechanical improvement in assembling, yet regardless it appears just as if machines are asserting authority over the business. Subsequently, it is fundamental to also investigate this framework in order to ensure no weakness make finishes on the monetary issues of work later on. This will help workers of today plan for a not so distant future.

Given the credibility of the business, it will show new employments in monstrous data evaluation, robot experts and a tremendous piece of mechanical coordinators. While attempting to pick the kind of occupations that Industry 4.0 will offer or need more labour in, B.C.G. has passed on a report based on get-togethers with 20 of the business' experts to show how ten of the biggest use cases for the foundation of the business will be influenced. The following is a selection of the critical changes that will impact the economics of work.

> **Big Data-Driven Quality Control:** In sorting out terms, quality control goes for decreasing the particular course of action between things. Quality control depends to an amazing degree on real systems to show whether a particular part of something (for instance, its size or weight) is changing in a way that can be seen, for instance. Regardless of what may be the normal side, the importance of Big Data experts will increase.
>
> **Robot-Assisted Production:** The entire explanation of the new business depends on the smart devices having the option to interface with the including condition. This proposes experts who help manufacturing (for instance, packaging) will be laid off and be replaced with smart devices outfitted with cameras, sensors and actuators that can disengage the thing and a brief time span later pass on the essential changes for it. Thusly, the need for such specialists will drop and they will be displaced with 'robot facilitators'.
>
> **Self-Driving Vehicles Logistics:** Chief among the most basic central explanations behind improvement is transportation. Manufacturers use straight programming procedures (for instance, the Transportation Model) to make use of transportation. Notwithstanding, with self-driving vehicles, and with the assistance of Big Data, such incalculable will be laid off.
>
> **Age Line Simulation:** While the basic for redesign for transportation rots, the requirement for modern manufacturers (who regularly manage streamlining and re-enactment) to recreate signs lines will broaden. Having the improvement to copy age lines before establishment will open up occupations for mechanical coordinators put huge imperativeness in the propelled field.

Perceptive Maintenance: Having mind blowing devices will draw in producers to imagine disappointments. Fast machines will presumably similarly manage themselves. Along these lines, the proportion of standard help masters with willing drop, and they'll be removed with fundamentally more in conviction taught ones.

Machines as a Service: The new business will likewise connect with clients to sell a machine as a membership. This means that as opposed to selling the entire machine to the client, the machine will be set up and maintained by the manufacturer while the client manhandles the associations it gives.

BIBLIOGRAPHY

1. Pozdnyakova, U.A., Golikov, V.V., Peters, I.A. and Morozova, I.A., 2019. Genesis of the revolutionary transition to industry 4.0 in the 21st century and overview of previous industrial revolutions. In *Industry 4.0: Industrial Revolution of the 21st Century* (pp. 11–19). Springer, Cham.
2. Dolgui, A., Ivanov, D., Sethi, S.P. and Sokolov, B., 2019. Scheduling in production, supply chain and Industry 4.0 systems by optimal control: fundamentals, state-of- the-art and applications. *International Journal of Production Research*, *57*(2), pp.411–432.
3. Manavalan, E. and Jayakrishna, K., 2019. A review of Internet of Things (IoT) embedded sustainable supply chain for industry 4.0 requirements. *Computers & Industrial Engineering*, *127*, pp.925–953.
4. Pace, P., Aloi, G., Gravina, R., Caliciuri, G., Fortino, G. and Liotta, A., 2018. An edge-based architecture to support efficient applications for healthcare Industry 4.0. *IEEE Transactions on Industrial Informatics*, *15*(1), pp.481–489.
5. Pace, P., Aloi, G., Gravina, R., Caliciuri, G., Fortino, G. and Liotta, A., 2018. An edge-based architecture to support efficient applications for healthcare Industry 4.0. *IEEE Transactions on Industrial Informatics*, *15*(1), pp.481–489.
6. Fantini, P., Pinzone, M. and Taisch, M., 2020. Placing the operator at the centre of Industry 4.0 design: modelling and assessing human activities within cyber-physical systems. *Computers & Industrial Engineering*, *139*, 105058.
7. Kugler, Z., Szabó, G., Abdulmuttalib, H.M., Batini, C., Shen, H., Barsi, A. and Huang, G., 2018. Time-related quality dimensions of urban remotely sensed big data. *International Archives of the Photogrammetry, Remote Sensing & Spatial Information Sciences, 624,* 315–320.
8. Uhlemann, T.H.J., Lehmann, C. and Steinhilper, R., 2017. The digital twin: realizing the cyber-physical production system for Industry 4.0. *Procedia CIRP, 61,* pp.335–340.
9. Witkowski, K., 2017. Internet of things, big data, industry 4.0–innovative solutions in logistics and supply chains management. *Procedia Engineering, 182,* pp.763–769.
10. Thoben, K.D., Wiesner, S. and Wuest, T., 2017. "Industrie 4.0" and smart manufacturing-a review of research issues and application examples. *International Journal of Automation Technology, 11*(1), pp.4–16.
11. Witkowski, K., 2017. Internet of things, big data, industry 4.0–innovative solutions in logistics and supply chains management. *Procedia Engineering, 182,* pp.763–769.
12. Simons, S., Abé, P. and Neser, S., 2017. Learning in the AutFab–the fully automated Industrie 4.0 learning factory of the University of Applied Sciences Darmstadt. *Procedia Manufacturing, 9,* pp.81–88.
13. Marilungo, E., Papetti, A., Germani, M. and Peruzzini, M., 2017. From PSS to CPS design: a real industrial use case toward Industry 4.0. *Procedia CIRP, 64,* pp.357–362.

14. Batista, N.C., Melício, R. and Mendes, V.M.F., 2017. Services enabler architecture for smart grid and smart living services providers under Industry 4.0. *Energy and Buildings*, *141*, pp.16–27.
15. Akaev, A., Sarygulov, A., and Sokolov, V. 2018. Digital economy: backgrounds, main drivers and new challenges. In *SHS Web of Conferences* (Vol. 44, p. 00006). EDP Sciences.
16. Hofmann, E. and Rüsch, M., 2017. Industry 4.0 and the current status as well as future prospects on logistics. *Computers in Industry*, *89*, pp.23–34.
17. Albertin, A.L. and de Moura Albertin, R.M., 2017. A internet das coisas irá muito além as coisas. *GV EXECUTIVO*, *16*(2), pp.12–17.
18. Mrugalska, B. and Wyrwicka, M.K., 2017. Towards lean production in Industry 4.0. *Procedia Engineering*, *182*, pp.466–473.
19. Theorin, A., Bengtsson, K., Provost, J., Lieder, M., Johnsson, C., Lundholm, T. and Lennartson, B., 2017. An event- driven manufacturing information system architecture for Industry 4.0. *International Journal of Production Research*, *55*(5), pp.1297–1311.
20. Trappey, A.J., Trappey, C.V., Govindarajan, U.H., Chuang, A.C. and Sun, J.J., 2017. A review of essential standards and patent landscapes for the Internet of Things: a key enabler for Industry 4.0. *Advanced Engineering Informatics*, *33*, pp.208–229.
21. Oesterreich, T.D. and Teuteberg, F., 2016. Understanding the implications of digitisation and automation in the context of Industry 4.0: a triangulation approach and elements of a research agenda for the construction industry. *Computers in Industry*, *83*, pp.121–139.
22. Qin, J., Liu, Y. and Grosvenor, R., 2016. A categorical framework of manufacturing for industry 4.0 and beyond. *Procedia CIRP*, *52*, pp.173–178.
23. Erol, S., Jäger, A., Hold, P., Ott, K. and Sihn, W., 2016. Tangible Industry 4.0: a scenario-based approach to learning for the future of production. *Procedia CIRP*, *54*(1), pp.13–18.
24. Neugebauer, R., Hippmann, S., Leis, M. and Landherr, M., 2016. Industrie 4.0-From the perspective of applied research. *Procedia CIRP*, *57*(1), pp.2–7.
25. Sipsas, K., Alexopoulos, K., Xanthakis, V. and Chryssolouris, G., 2016. Collaborative maintenance in flow-line manufacturing environments: an Industry 4.0 approach. *Procedia CIRP*, *55*, pp.236–241.
26. Lee, J., Kao, H.A. and Yang, S., 2014. Service innovation and smart analytics for industry 4.0 and big data environment. *Procedia CIRP*, *16*, pp.3–8.
27. Schumacher, A., Erol, S. and Sihn, W., 2016. A maturity model for assessing Industry 4.0 readiness and maturity of manufacturing enterprises. *Procedia CIRP*, *52*, pp.161–166.
28. Dutra, D.D.S. and Silva, J.R., 2016. Product-Service Architecture (PSA): toward a service engineering perspective in Industry 4.0. *IFAC-PapersOnLine*, *49*(31), pp.91–96.
29. Landherr, M., Schneider, U. and Bauernhansl, T., 2016. The Application Center Industrie 4.0-Industry-driven manufacturing, research and development. *Procedia CIRP*, *57*, pp.26–31.
30. Ivanov, D., Sokolov, B. and Ivanova, M., 2016. Schedule coordination in cyber-physical supply networks Industry 4.0. *IFAC-PapersOnLine*, *49*(12), pp.839–844.
31. Wang, S., Wan, J., Li, D. and Zhang, C., 2016. Implementing smart factory of industrie 4.0: an outlook. *International Journal of Distributed Sensor Networks*, *12*(1), p.3159805.
32. Albers, A., Gladysz, B., Pinner, T., Butenko, V. and Stürmlinger, T., 2016. Procedure for defining the system of objectives in the initial phase of an industry 4.0 project focusing on intelligent quality control systems. *Procedia CIRP*, *52*, pp.262–267.
33. Gilchrist, A., 2016. *Industry 4.0: The Industrial Internet of Things*. Apress.
34. Grangel-González, I., Halilaj, L., Coskun, G., Auer, S., Collarana, D. and Hoffmeister, M., 2016, February. Towards a semantic administrative shell for industry 4.0 components. In *2016 IEEE Tenth International Conference on Semantic Computing (ICSC)* (pp. 230–237). IEEE.

35. Hecklau, F., Galeitzke, M., Flachs, S. and Kohl, H., 2016. Holistic approach for human resource management in Industry 4.0. *Procedia CIRP*, *54*, pp.1–6.

36. Hermann, M., Pentek, T. and Otto, B., 2016, January. Design principles for industrie 4.0 scenarios. In *2016 49th Hawaii International Conference on System Sciences (HICSS)* (pp. 3928–3937). IEEE.

37. Herter, J. and Ovtcharova, J., 2016. A model based visualization framework for cross discipline collaboration in Industry 4.0 scenarios. *Procedia CIRP*, *57*, pp.398–403.

38. Sipsas, K., Alexopoulos, K., Xanthakis, V. and Chryssolouris, G., 2016. Collaborative maintenance in flow-line manufacturing environments: an Industry 4.0 approach. *Procedia CIRP*, *55*, pp.236–241.

39. Stock, T. and Seliger, G., 2016. Opportunities of sustainable manufacturing in Industry 4.0. *Procedia CIRP*, *40*, pp.536–541.

40. Thames, L. and Schaefer, D., 2016. Software-defined cloud manufacturing for Industry 4.0. *Procedia CIRP*, *52*, pp.12–17.

41. Bahrin, M.A.K., Othman, M.F., Azli, N.N. and Talib, M.F., 2016. Industry 4.0: a review on industrial automation and robotic. *Jurnal Teknologi*, *78*(6–13), pp.137–143.

42. Rennung, F., Luminosu, C.T. and Draghici, A., 2016. Service provision in the framework of Industry 4.0. *Procedia-Social and Behavioral Sciences*, *221*, pp.372–377.

43. T Stock, T. and Seliger, G., 2016. Opportunities of sustainable manufacturing in Industry 4.0. *Procedia CIRP*, *40*, pp.536–541.

44. Erol, S., Jäger, A., Hold, P., Ott, K. and Sihn, W., 2016. Tangible Industry 4.0: a scenario-based approach to learning for the future of production. *Procedia CIRP*, *54*, pp.13–18.

45. Vijaykumar, S., Saravanakumar, S.G., & Balamurugan, M. (2016). Unique sense: smart computing prototype for industry 4.0 revolution with IOT and bigdata implementation model. *arXiv preprint arXiv:1612.09325*.

46. Wan, J., Tang, S., Shu, Z., Li, D., Wang, S., Imran, M. and Vasilakos, A.V., 2016. Software-defined industrial internet of things in the context of Industry 4.0. *IEEE Sensors Journal*, *16*(20), pp.7373–7380.

47. Wang, K., 2016. Intelligent predictive maintenance (IPdM) system–Industry 4.0 scenario. *WIT Transactions on Engineering Sciences*, *113*, pp.259–268.

48. Wang, S., Wan, J., Zhang, D., Li, D. and Zhang, C., 2016. Towards smart factory for industry 4.0: a self-organized multi- agent system with big data based feedback and coordination. *Computer Networks*, *101*, pp.158–168.

49. Weyer, S., Schmitt, M., Ohmer, M. and Gorecky, D., 2015. Towards Industry 4.0-Standardization as the crucial challenge for highly modular, multi-vendor production systems. *Ifac-Papersonline*, *48*(3), pp.579–584.

50. Faller, C. and Feldmüller, D., 2015. Industry 4.0 learning factory for regional SMEs. *Procedia CIRP*, *32*, pp.88–91.

51. Long, F., Zeiler, P. and Bertsche, B., 2016. Modelling the production systems in industry 4.0 and their availability with high-level Petri nets. *IFAC-PapersOnLine*, *49*(12), pp.145–150.

52. Long, F., Zeiler, P. and Bertsche, B., 2016. Modelling the production systems in industry 4.0 and their availability with high-level Petri nets. *IFAC-PapersOnLine*, *49*(12), pp.145–150.

53. Qin, J., Liu, Y. and Grosvenor, R., 2016. A categorical framework of manufacturing for industry 4.0 and beyond. *Procedia CIRP*, *52*, pp.173–178.

54. Rennung, F., Luminosu, C.T. and Draghici, A., 2016. Service provision in the framework of Industry 4.0. *Procedia-Social and Behavioral Sciences*, *221*, pp.372–377.

55. Wittenberg, C., 2016. Human-CPS Interaction-requirements and human-machine interaction methods for the Industry 4.0. *IFAC-PapersOnLine*, *49*(19), pp.420–425.

56. Zezulka, F., Marcon, P., Vesely, I. and Sajdl, O., 2016. Industry 4.0–An Introduction in the phenomenon. *IFAC- PapersOnLine, 49*(25), pp.8–12.
57. Zhou, K., Liu, T. and Zhou, L., 2015, August. Industry 4.0: towards future industrial opportunities and challenges. In *2015 12th International Conference on Fuzzy Systems and Knowledge Discovery (FSKD)* (pp. 2147–2152). IEEE.
58. Saldivar, A.A.F., Li, Y., Chen, W.N., Zhan, Z.H., Zhang, J. and Chen, L.Y., 2015, September. Industry 4.0 with cyber-physical integration: a design and manufacture perspective. In *2015 21st International Conference on Automation and Computing (ICAC)* (pp. 1–6). IEEE.
59. Almada-Lobo, F., 2015. The Industry 4.0 revolution and the future of Manufacturing Execution Systems (MES). *Journal of Innovation Management, 3*(4), pp.16–21.
60. Bagheri, B., Yang, S., Kao, H.A. and Lee, J., 2015. Cyber-physical systems architecture for self-aware machines in industry 4.0 environment. *IFAC-PapersOnLine, 48*(3), pp.1622–1627.
61. Haddara, M. and Elragal, A., 2015. The readiness of ERP systems for the factory of the future. *Procedia Computer Science, 64*, pp.721–728.
62. Valdeza, A.C., Braunera, P., Schaara, A.K., Holzingerb, A. and Zieflea, M., 2015, August. Reducing complexity with simplicity-usability methods for Industry 4.0. In *Proceedings 19th Triennial Congress of the IEA* (Vol. 9, p. 14).
63. Kolberg, D. and Zühlke, D., 2015. Lean automation enabled by industry 4.0 technologies. *IFAC-PapersOnLine, 48*(3), pp.1870–1875.
64. Schumacher, A., Erol, S. and Sihn, W., 2016. A maturity model for assessing Industry 4.0 readiness and maturity of manufacturing enterprises. *Procedia CIRP, 52*, pp.161–166.
65. Schuster, K., Groß, K., Vossen, R., Richert, A. and Jeschke, S., 2016. Preparing for industry 4.0–collaborative virtual learning environments in engineering education. In *Engineering Education 4.0* (pp. 477–487). Springer, Cham.
66. Shafiq, S.I., Sanin, C., Szczerbicki, E. and Toro, C., 2016. Virtual engineering factory: creating experience base for Industry 4.0. *Cybernetics and Systems, 47*(1–2), pp.32–47.
67. Shafiq, S.I., Sanin, C., Toro, C. and Szczerbicki, E., 2015. Virtual engineering object (VEO): toward experience-based design and manufacturing for Industry 4.0. *Cybernetics and Systems, 46*(1–2), pp.35–50.
68. Ivanov, D., Sokolov, B. and Ivanova, M., 2016. Schedule coordination in cyber-physical supply networks Industry 4.0. *IFAC-PapersOnLine, 49*(12), pp.839–844.
69. Kolberg, D. and Zühlke, D., 2015. Lean automation enabled by industry 4.0 technologies. *IFAC-PapersOnLine, 48*(3), pp.1870–1875.
70. Weyer, S., Schmitt, M., Ohmer, M. and Gorecky, D., 2015. Towards Industry 4.0-Standardization as the crucial challenge for highly modular, multi-vendor production systems. *Ifac-Papersonline, 48*(3), pp.579–584.
71. Sommer, L., 2015. Industrial revolution-industry 4.0: are German manufacturing SMEs the first victims of this revolution? *Journal of Industrial Engineering and Management, 8*(5), pp.1512–1532.
72. Rüßmann, M., Lorenz, M., Gerbert, P., Waldner, M., Justus, J., Engel, P. and Harnisch, M., 2015. Industry 4.0: the future of productivity and growth in manufacturing industries. *Boston Consulting Group, 9*(1), pp.54–89.
73. Gölzer, P., Cato, P. and Amberg, M., 2015, May. Data processing requirements of industry 4.0-use cases for big data applications. In *ECIS*.
74. Pan, M., Sikorski, J., Kastner, C.A., Akroyd, J., Mosbach, S., Lau, R. and Kraft, M., 2015. Applying Industry 4.0 to the Jurong Island eco-industrial park. *Energy Procedia, 75*, pp.1536–1541.
75. Veza, I., Mladineo, M. and Gjeldum, N., 2015. Managing innovative production network of smart factories. *IFAC-PapersOnLine, 48*(3), pp.555–560.

76. Schmidt, R., Möhring, M., Härting, R.C., Reichstein, C., Neumaier, P. and Jozinović, P., 2015, June. Industry 4.0- potentials for creating smart products: empirical research results. In *International Conference on Business Information Systems* (pp. 16–27). Springer, Cham.

77. Schuh, G., Gartzen, T., Rodenhauser, T. and Marks, A., 2015. Promoting work-based learning through Industry 4.0. *Procedia CIRP, 32,* pp.82–87.

78. Schuh, G., Potente, T., Wesch-Potente, C., Weber, A.R. and Prote, J.P., 2014. Collaboration Mechanisms to increase Productivity in the Context of Industrie 4.0. *Procedia CIRP, 19,* pp.51–56.

79. Valdeza, A.C., Braunera, P., Schaara, A.K., Holzingerb, A. and Zieflea, M., 2015, August. Reducing complexity with simplicity-usability methods for Industry 4.0. In *Proceedings 19th Triennial Congress of the IEA* (Vol. 9, p. 14).

80. Brettel, M., Friederichsen, N., Keller, M. and Rosenberg, M., 2014. How virtualization, decentralization and network building change the manufacturing landscape: an Industry 4.0 perspective. *International Journal of Mechanical, Industrial Science and Engineering, 8*(1), pp.37–44.

81. Schuh, G., Potente, T., Wesch-Potente, C., Weber, A.R. and Prote, J.P., 2014. Collaboration Mechanisms to increase Productivity in the Context of Industrie 4.0. *Procedia CIRP, 19,* pp.51–56.

82. Gorecky, D., Schmitt, M., Loskyll, M. and Zühlke, D., 2014, July. Human-machine-interaction in the industry 4.0 era. In *2014 12th IEEE International Conference on Industrial Informatics (INDIN)* (pp. 289–294). IEEE.

83. Lee, J., Bagheri, B. and Kao, H.A., 2015. A cyber-physical systems architecture for industry 4.0-based manufacturing systems. *Manufacturing Letters, 3,* pp.18–23.

84. Paelke, V., 2014, September. Augmented reality in the smart factory: supporting workers in an Industry 4.0. environment. In *Proceedings of the 2014 IEEE Emerging Technology and Factory Automation (ETFA)* (pp. 1–4). IEEE.

85. Shrouf, F., Ordieres, J. and Miragliotta, G., 2014, December. Smart factories in Industry 4.0: a review of the concept and of energy management approached in production based on the Internet of Things paradigm. In *2014 IEEE International Conference on Industrial Engineering and Engineering Management* (pp. 697–701). IEEE.

86. Lee, J., Kao, H.A. and Yang, S., 2014. Service innovation and smart analytics for industry 4.0 and big data environment. *Procedia CIRP, 16,* pp.3–8.

87. Brettel, M., Friederichsen, N., Keller, M. and Rosenberg, M., 2014. How virtualization, decentralization and network building change the manufacturing landscape: an Industry 4.0 perspective. *International Journal of Mechanical, Industrial Science and Engineering, 8*(1), pp.37–44.

88. Imtiaz, J. and Jasperneite, J., 2013, July. Scalability of OPC-UA down to the chip level enables "Internet of Things". In *2013 11th IEEE International Conference on Industrial Informatics (INDIN)* (pp. 500–505). IEEE.

89. Gubbi, J., Buyya, R., Marusic, S. and Palaniswami, M., 2013. Internet of Things (IoT): a vision, architectural elements, and future directions. *Future Generation Computer Systems, 29*(7), pp.1645–1660.

90. Gluhak, A., Krco, S., Nati, M., Pfisterer, D., Mitton, N. and Razafindralambo, T., 2011. A survey on facilities for experimental internet of things research.

91. Givehchi, O., Trsek, H. and Jasperneite, J., 2013, September. Cloud computing for industrial automation systems—A comprehensive overview. In *2013 IEEE 18th Conference on Emerging Technologies & Factory Automation (ETFA)* (pp. 1–4). IEEE.

92. Weber, R.H., 2010. Internet of Things–New security and privacy challenges. *Computer Law & Security Review, 26*(1), pp.23–30.

93. Atziori, L., Iera, A. and Morabito, G., 2010. The Internet of Things: a survey computer networks. 54(15), 2787–2805.
94. Kitchenham, B., Brereton, O.P., Budgen, D., Turner, M., Bailey, J. and Linkman, S., 2009. Systematic literature reviews in software engineering–a systematic literature review. *Information and Software Technology*, *51*(1), pp.7–15.
95. Hozdić, E., 2015. Smart factory for industry 4.0: a review. *International Journal of Modern Manufacturing Technologies*, *7*(1), pp.28–35.
96. Jazdi, N., 2014, May. Cyber physical systems in the context of Industry 4.0. In *2014 IEEE International Conference on Automation, Quality and Testing, Robotics* (pp. 1–4). IEEE.

3.4 ANALYSIS OF DRIVERS FOR CLOUD MANUFACTURING AND ITS INTEGRATION WITH INDUSTRY 4.0 USING THE MCDM TECHNIQUE

S. Vinodh and Vishal A. Wankhede

3.4.1 INTRODUCTION

Manufacturing organisations are moving towards the fourth industrial revolution (Industry 4.0). Cloud manufacturing is a customer-focused manufacturing approach that makes use of on-demand distributed and diversified manufacturing resources to increase effectiveness of reconstructing production lines, decrease product life-cycle costs and ensure minimal use of resources with respect to variable customer demand (Wu et al., 2013). Industry 4.0 was primarily projected in 2011 for German economic development (Roblek et al., 2016). New developments in industry have continued to exist for more than a few hundred years and at present, it is in the period of Industry 4.0. The fourth industrial revolution i.e., Industry 4.0, exploits cyber-physical systems and cloud-based concepts in the manufacturing sector (Liu et al., 2017). In order to deal with the manufacturing industry's challenges like increasing productivity, quick response to customer needs, workforce performance, etc. there is a need to adopt information technology (I.T.) solutions. Use of these technologies helps industry to compete with markets and continue this worldwide. Moreover, analysing the drivers for cloud manufacturing and Industry 4.0 integration aids effective implementation in industry. Industry 4.0 includes I.o.T., cyber-physical systems, cloud systems, automation and so on. Cloud manufacturing is a vital driver of Industry 4.0. In this viewpoint, this chapter presents the analysis of drivers for cloud manufacturing amalgamated with Industry 4.0. 20 drivers are analysed using the M.C.D.M. (Multi-Criteria Decision-Making) tool A.H.P. (Analytical Hierarchy Process). The priority order of drivers is derived. The practical inferences are presented.

3.4.2 LITERATURE REVIEW

Castelo-Branco et al. (2019) studied the factors influencing the implementation of Industry 4.0 in manufacturing organisations across European countries. They adopted factor analysis and cluster analysis tools for their study purposes. The study revealed that integration of digital infrastructure and analytical support to cope with huge data demonstrates the inclination for Industry 4.0. Turkes et al. (2019) identified the barriers and drivers to using Industry 4.0 through a constructive sampling method. Research methodology includes questionnaire circulated to 176 managers of S.M.E.s in Romania. Collected responses were analysed using SPSS. The finding showed a lack of knowledge about Industry 4.0 found to be the most influential driver and barrier towards using Industry 4.0. Additive manufacturing and augmented reality provide improved methods to arrange maintenance operations rather than traditional methods. Kamble et al. (2018) analysed the key barriers among 12 barriers that stop manufacturing organisations from accepting Industry 4.0. They developed interrelationships between barriers with the help of interpretative structural modelling. Moreover, they found dependence and the driving power of barriers by means of MICMAC analysis. It is apparent from the study that 'legal and contractual uncertainty' barrier was found to be the most influential. Liu et al. (2017) showed comparative analysis of cloud manufacturing and Industry 4.0 from different point of view. Findings include Industry 4.0 to be a more widespread concept than cloud manufacturing as it considers both horizontal and vertical integration, whereas cloud manufacturing focuses only on cloud integration. Wu et al. (2013) presented various drivers associated with cloud manufacturing and discussed its present status and future scope. They also discussed the influence of long- and short-term impacts of cloud manufacturing on various streams. In their study, they concluded that cloud manufacturing would facilitate manufacturing organisations to construct open and virtual manufacturing processes. Also, it allows different stakeholders to do business more easily and cost effectively. Vaidya et al. (2018) demonstrated a glimpse of Industry 4.0 and applications of its nine pillars. They also identified challenges and problems associated with its implementation. They suggested new research avenues to be discovered like industrial management and transparent supply chains. Yang Lu (2017) projected the vital issue of Industry 4.0 interoperability and presented a conceptual framework with challenges and future trends in Industry 4.0 research. They conducted a literature review by following the two-state approach identified by Watson and Webster. The study found time- and cost-efficiency increases from the implementation of Industry 4.0, along with product quality improvement. Carvalho et al. (2018) described the linkage of Industry 4.0 with sustainability. The study showed the advantages of the novel industry model, like enhanced product life-cycles, virtualisation, decentralisation and interoperability which lead to further adaptability for available natural resources and environmental costs. Thames et al. (2016) introduced the idea of software-defined cloud manufacturing which bridges the gap between hardware and cloud-based applications and platforms. The aim of this integration was to advance cloud manufacturing and Industry 4.0 pillars by providing adaptability and agility and thereby reducing complexity. Shankar et al. (2016) analysed advanced sustainable manufacturing system drivers using the Analytical Hierarchy Process (A.H.P.). The aim of the study was to amalgamate advanced manufacturing methods with sustainable operations. Findings showed that quality was the key driver that influences manufacturing organisation to adopt advanced sustainable manufacturing. Vinodh et al. (2011) presented selection of lean concepts in the manufacturing industry using A.H.P. methodology. Stock et al. (2018) studied Industry 4.0 as a potential enabler for sustainable development through qualitative assessment of three steps. The evaluation of social and ecological aspects was done using the Delphi method. The findings revealed that value creation was the most influential factor contributing to the environmental element of sustainability. Rubmann et al. (2015) discussed nine technology elements of Industry 4.0 and shared its economic and technical advantages for production and manufacturing product suppliers.

3.4.3 METHODOLOGY

Multi-Criteria Decision-Making (M.C.D.M.) techniques are used to deal with multiple criteria-associated problems. Various M.C.D.M. tools have been proposed by researchers, but A.H.P. has been found to be most important. A.H.P. handles various complex M.C.D.M. problems which makes it reliable in a new environment. In 1980, A.H.P. was developed by Prof. Thomas Saaty and was applied in various fields with the help of expert decision-makers. A.H.P. models a hierarchy structure to determine the influence from one stage to another. A.H.P. also handles multiple factors that are used in decision-making. A.H.P. has been successfully applied in the manufacturing sector, as it allows experts to make certain of the consistency of their decision regarding the criteria.

The stages of the A.H.P. methodology are given below (Yadav et al., 2016):

1. Collect various drivers pertaining to cloud manufacturing and its integration with Industry 4.0 from the literature and from subject experts.
2. Prepare an input sheet and create a pair-wise comparison of the criteria with the help of expert opinion. This assessment will be based on the Saaty scale (Table 3.2).
3. Prioritisation weight calculations using normalisation results.
4. Validating the expert responses using consistency check of the matrices.
5. For consistency, it is expected to have a consistency index of less than 10% (0.1), or up to 20% can be accepted in some cases, but not more than that. If the verdict is not satisfactory, then experts are asked to revise the input sheet to explain the error.
6. The drivers of cloud manufacturing and its integration with Industry 4.0 are finally prioritised using the final weight calculation.

TABLE 3.2
Saaty Scale of Relative Importance

Importance	Description
9	Extreme
7	Very strong
5	Strong
3	Moderate
1	Equal
Even numbers	Intermediate

3.4.4 CASE STUDY

The study has been conducted to analyse the drivers pertaining to cloud manufacturing and its integration with Industry 4.0. A.H.P. methodology, an M.C.D.M. tool, is selected for the analysis. Table 3.3 shows 20 drivers which have been identified from the literature research and discussion with subject experts.

3.4.5 ANALYSIS USING A.H.P. METHODOLOGY

The procedure is explained below:

1. Pair-wise comparison matrix from responses of decision experts from industry as well as academia (Table 3.4).
2. Normalisation of pair-wise comparison was made using mathematical calculations.
3. Normalised data is further examined for consistency check using Eigen values by ensuring a Consistency Ratio (C.R.) of less than 0.1.
4. Lastly, drivers are ranked based on the Eigen values derived for each driver (Table 3.7).

3.4.6 NORMALISATION CALCULATION

For Driver D2:

1. Multiplication: $3\times1\times3\times3\times3\times3\times3\times3\times5\times3\times3\times3\times3\times1/3\times5\times1\times3\times3\times3\times3$ = **119574225**
2. $(\text{Multiplication})^{\frac{1}{\text{Matrix index}}}$; $(119574225)^{1/20}$ = **2.534439**
3. Weight = (2.534439)/Weight summation = 2.534439/23.3208 = **0.108667**

Thus, the weights for each driver are calculated (Table 3.5) and ranked accordingly. Further the matrix was checked for its consistency by computing the Consistency Index and the Consistency Ratio using Eq. (3.5.1) and Eq. (3.5.2).

Table 3.6 shows the consistency calculation for the pair-wise matrix.

$$\text{Consistency Index}\left(\text{CI}\right) = \left(\text{Sumtotal} - N\right)/\left(N - 1\right) \qquad (3.5.1)$$

Where N= Size of the matrix
CI = (24.266 – 20)/ 19 = 0.224
Random Index (RI) = 1.62 (for N = 20)

Thus,

$$\text{Consistency Ratio}\left(\text{CR}\right) = \text{CI/RI}$$
$$= 0.224/1.62 = \textbf{0.138}$$
$$(3.5.2)$$

TABLE 3.3

Drivers of Cloud Manufacturing and Industry 4.0 Integration and Its Description

Sr. No.	Drivers	Description	References
1.	Cloud computing (D1)	Cloud computing is storing and accessing data and programs over the internet instead of computer's hard drive which provides software as a service to everyone. This breaks the walls of location barriers to accessing resources.	Yang Lu (2017), Nubia Carvalho (2018)
2.	Economy (D2)	Acceptance of cloud manufacturing by organisations is due to promising crowd-sourcing and outsourcing models in manufacturing and design. These models aid small–medium enterprises to decrease costs by outsourcing their support services and operations.	Wu et al. (2013)
3.	Knowledge management systems (D3)	Handling the real-time data is need to Industry 4.0 which can be done by understanding. It ensures storing and retrieving knowledge of data to the systems.	Kamble et al. (2018)
4.	Internet coverage and IT facilities (D4)	IT infrastructure that is required to support Industry 4.0 implementation.	Kamble et al. (2018)
5.	Artificial Intelligence (D5)	Artificial Intelligence is nothing but the mimic of human intelligence. It bridges the gap between implementation of Industry 4.0 and cloud manufacturing.	Kamble et al. (2018)
6.	Agility (D6)	Agility is a distinctive driver of cloud manufacturing and Industry 4.0 which allows redesigning of products and related manufacturing aspects to meet quick-changing customer demands.	Wu et al. (2013), Lane Thames et al. (2016)
7.	Scalability (D7)	Scalability ensures flexibility in cloud manufacturing and Industry 4.0 environment. Cloud manufacturing permits organisations to rapidly scale the enterprise structure i.e., machines, machine tools, handling units, employers, etc. can be tailored as per the requirements quickly.	Wu et al. (2013)
8.	Resource sharing and collaboration (D8)	Manufacturing resource-sharing and collaboration between service providers enhance cloud manufacturing environment.	Wu et al. (2013)
9.	Industrial Internet of Things (D9)	Industry 4.0 is noticeable by high automation along with the use of information technology and electronics, like the internet of things in the manufacturing industry.	Yang Lu (2017), Nubia Carvalho (2018)
10.	Enterprise architecture (D10)	For effective implementation of Industry 4.0, enterprise architecture provides a platform for cloud, predictive analytics, digital technologies and system interfaces.	Yang Lu (2017)
11.	Enterprise integration (D11)	Enterprise integration ensures industry keeps the existing system up-to-date.	Yang Lu (2017)

(Continued)

TABLE 3.3 (CONTINUED)

Drivers of Cloud Manufacturing and Industry 4.0 Integration and Its

Sr. No.	Drivers	Description	References
12.	Smart factory (D12)	Transformation of traditional system to fully coupled and flexible automation defines smart factory. Industry 4.0 integrates smart factory and artificial intelligence.	Yongkui Liu et al. (2016)
13.	Information sharing (D13)	Huge data with unstructured details of the manufacturing process and product design are stored in information and communication technology systems. Cloud manufacturing ensures proper processing, managing and sharing of this information across and within industry with the cloud manufacturing domain.	Wu et al. (2013)
14.	Social sphere (D14)	Developing social linkage over the internet between various organisations for effective communication in cloud manufacturing and the Industry 4.0 environment.	Wu et al. (2013)
15.	Cyber-physical systems (D15)	A cyber-physical system enhances growth, productivity, workforce performance and producing low-cost good quality products by collecting and analysing harmful data.	Yang Lu (2017), Nubia Carvalho (2018)
16.	Top management support (D16)	Support extended by the higher-ups for successful integration of cloud manufacturing and Industry 4.0.	(Lee et al., 2007)
17.	Firm size (D17)	Organisation volume influences the acceptance of integration technology.	(Zang et al., 2014)
18.	Technological readiness (D18)	Assessment method to determine acceptance level of particular technology. Cloud manufacturing and Industry 4.0 integration as a technology; its maturity can be evaluated by technological readiness.	(Oliveira et al., 2014)
19.	Relative advantage (D19)	The degree to which an innovation is perceived as being better than the idea it supersedes.	(Oliveira et al., 2014)
20.	Complexity (D20)	Complexity is the extent to which an improvement is professed as comparatively complicated to recognise and utilise.	(Wang et al., 2017)

3.4.7 RESULTS AND DISCUSSION

The study presents analysis of drivers for cloud manufacturing and its integration with Industry 4.0 using Analytical Hierarchy Process (A.H.P.). Use of this integration helps industry to compete with markets and sustain worldwide. In this viewpoint, analysing the drivers for cloud manufacturing and Industry 4.0 aids effective implementation in industry. Table 3.7 reveals social sphere (D14) as the most influential driver with weight of 12.1% (0.121765), followed by the economy (D2) driver, 10.8%, and cyber-physical systems (D20), 1.93% (Table 3.7).

TABLE 3.4

Pair-Wise Comparison Matrix (P.W.C.M.) of Drivers of Cloud Manufacturing and Industry 4.0 Integration

	D1	D2	D3	D4	D5	D6	D7	D8	D9	D10	D11	D12	D13	D14	D15	D16	D17	D18	D19	D20
D1	1	1/3	1	1/3	1	1	1/3	1/3	3	1/3	1/3	3	1/3	1/3	1	1	1/3	1	1/3	1/3
D2	3	1	3	3	3	3	3	3	5	3	3	3	3	1/3	5	1	3	3	3	3
D3	1	1/3	1	1/3	3	1/3	1/3	3	3	3	3	3	1	1/3	1	1/3	3	1/3	1/3	3
D4	3	1/3	3	1	3	3	3	1	3	3	3	3	1	1/3	3	1/3	3	3	3	3
D5	1	1/3	1/3	1/3	1	1/3	1/3	1/3	1	1/3	3	1	1/3	1/3	1	1/3	1/3	1/3	1/3	1/3
D6	1	1/3	3	1/3	3	1	1/3	1/3	3	1/3	3	3	3	1/3	3	1/3	3	1/3	3	1/3
D7	3	1/3	3	1/3	3	3	1	3	3	3	3	3	3	1/3	3	1	3	3	1	3
D8	3	1/3	1/3	1	3	3	1/3	1	3	3	3	3	3	1/3	3	3	3	3	3	3
D9	1/3	1/5	1/3	1/3	1	1/3	1/3	1/3	1	1/3	1/3	1	1	1/3	1	1/3	3	1	1/3	1/3
D10	3	1/3	1/3	1/3	3	1/3	1/3	1/3	3	1	1/3	1	1/3	1/3	3	1	1/3	3	3	3
D11	3	1/3	1/3	1/3	1/3	3	1/3	1/3	3	3	1	3	1/3	1/3	3	1	1/3	1/3	3	3
D12	1/3	1/3	1	1/3	1	1/3	1/3	1	1	1/3	1/3	1	1/3	1/3	1	1/3	1/3	1	1/3	3
D13	3	1/3	1	1	3	1/3	3	3	3	3	1	3	1	1/3	3	1/3	3	3	3	3
D14	3	3	3	3	3	3	3	3	3	3	3	3	3	1	3	3	3	3	3	3
D15	1	1/5	1	1/3	1	1/3	1/3	1/3	1	1/3	1/3	1	1/3	1/3	1	1/3	1/3	1/3	1/3	1/3
D16	1	1	3	3	3	3	1	1/3	3	1	3	3	3	1/3	3	1	3	3	3	3
D17	3	1/3	1/3	1/3	3	3	1/3	1/3	1/3	3	3	3	3	1/3	3	1/3	1	3	1	3
D18	1	1/3	3	1/3	3	3	1	1/3	1	3	1	1	1/3	1/3	3	3	1/3	1	1	1/3
D19	3	1/3	3	1/3	3	1/3	1/3	1/3	3	1/3	3	1	1/3	1/3	3	1/3	1/3	3	1	1/3
D20	3	1/3	1/3	1/3	3	3	1/3	1/3	3	3	3	3	3	1/3	3	1/3	1/3	3	3	1

TABLE 3.5

Normalised Matrix of Drivers of Cloud Manufacturing and Industry 4.0 Integration

	D1	D2	D3	D4	D5	D6	D7	D8	D9	D10	D11	D12	D13	D14	D15	D16	D17	D18	D19	D20	Multiplication	$(\text{Multiplication})^{\frac{1}{\text{Matrix index}}}$	Weights
D1	1	1/3	1	1/3	1	1	1/3	1/3	3	1/3	1/3	3	1/3	1/3	1	1	1/3	1	1/3	1/3	0	0.609952	0.026155
D2	3	1	3	3	3	3	3	3	5	3	3	3	3	1/3	5	1	3	3	3	3	119574225	2.534439	0.108677
D3	1	1/3	1	1/3	3	1/3	1/3	3	3	3	3	3	1	1/3	1	1/3	3	1/3	1/3	3	1	1	0.04288
D4	3	1/3	3	1	3	3	3	1	3	3	3	3	1	1/3	3	1/3	3	3	3	3	177147	1.829855	0.078465
D5	1	1/3	1/3	1/3	1	1/3	1/3	1/3	1	1/3	3	1	1/3	1/3	1	1/3	1/3	1/3	1/3	1/3	0	0.489634	0.020996
D6	1	1/3	3	1/3	3	1	1/3	1/3	3	1/3	3	3	3	1/3	3	1/3	3	3	1/3	1/3	1/9	0.895958	0.038419
D7	3	1/3	3	1/3	3	1	1	3	3	3	3	3	3	1/3	3	1	3	3	1	3	177147	1.829855	0.078465
D8	3	1/3	3	1	3	3	1/3	1	3	3	3	1	1	1/3	3	3	3	3	3	3	6561	1.551846	0.066543
D9	1/3	1/5	1/3	1/3	1	1/3	1/3	1/3	1	1/3	1/3	1	1/3	1/3	1	1/3	3	3	1/3	1/3	0	0.477286	0.020466
D10	3	1/3	1/3	1/3	3	3	1/3	1/3	3	1	1/3	3	1/3	1/3	3	1	3	1/3	3	1/3	0	0.802742	0.034422
D11	3	1/3	1/3	1/3	1/3	1/3	1/3	1	1	3	1	3	1	1/3	1	1/3	1/3	1	1/3	1/3	0	0.680781	0.029192
D12	1/3	1/3	1/3	1/3	1	1/3	1/3	1	1	1/3	1/3	1	1/3	1/3	1	1/3	1/3	1	1	1/3	0	0.489634	0.020996
D13	3	1/3	1	1	3	1/3	3	3	3	3	1	3	1	1/3	3	1/3	3	3	3	3	3	1.056467	0.045302
D14	3	3	3	3	3	1/3	3	3	3	3	3	3	3	1	3	3	3	3	3	3	1162261467	2.839652	0.121765
D15	1	1/5	1	1/3	1	1/3	1/3	1/3	3	1/3	1/3	1	1/3	1/3	1	1/3	1/3	1/3	1/3	1/3	0	0.451776	0.019372
D16	1	1	3	3	3	3	1	1/3	3	1	3	3	3	1/3	3	1	3	3	3	3	177147	1.829855	0.078465
D17	3	1/3	1/3	1/3	3	3	1/3	1/3	1	3	1	1	3	1/3	3	1/3	1	3	1	3	9	1.116123	0.04786
D18	1	1/3	3	1/3	3	3	1/3	1/3	1	3	1	1	1/3	1/3	3	1/3	1/3	1	1/3	1/3	0	0.759836	0.032582
D19	3	1/3	3	1/3	3	1/3	1	1/3	3	1/3	3	1	1/3	1/3	3	1/3	1	3	1	1/3	1/9	0.895958	0.038419
D20	3	1/3	1/3	1/3	3	3	1/3	1/3	3	3	3	3	3	1/3	3	1/3	1/3	3	3	1	27	1.179148	0.050562

TABLE 3.6
Consistency Calculations of Pair-Wise Comparison Matrix (P.W.C.M.)

Driver	P.W.C.M. × Weights	(P.W.C.M. × Weight)/Weight
D1	0.616476	23.5702676
D2	2.380687	21.9060362
D3	1.198568	27.9515747
D4	1.79563	22.884615
D5	0.483168	23.0128603
D6	1.010041	26.2902449
D7	1.785553	22.7561888
D8	1.639292	24.634922
D9	0.522743	25.5418697
D10	0.901449	26.1884069
D11	0.728478	24.9546996
D12	0.479582	22.8420703
D13	1.077339	23.7815289
D14	2.75647	22.6376589
D15	0.419419	21.6505623
D16	1.845479	23.5199188
D17	1.238842	25.8849475
D18	0.83583	25.6531978
D19	0.968206	25.2013345
D20	1.237226	24.4694481
	Sum Total	**24.2666176**

The ranking of drivers is as follows:

Social sphere (D14) > Economy (D2) > Internet coverage and IT Facilities (D4) > Scalability (D7) > Top management support (D16) > Resource sharing and collaboration (D8) > Complexity (D20) > Firm size (D17) > Information sharing (D13) > Knowledge management systems (D3) > Agility (D6) > Relative advantage (D19) > Enterprise architecture (D10) > Technological readiness (D18) > Enterprise integration (D11) > Cloud computing (D1) > Artificial Intelligence (D5) >S mart factory (D12) > Industrial internet of things (D9) > Cyber-physical systems (D15).

The prioritised drivers would enable industry practitioners to deal with challenges of Industry 4.0.

3.4.8 CONCLUSION

This article analyses the drivers of cloud manufacturing integrated with Industry 4.0. 20 drivers are identified and analysed. The analysis is formulated as an MCDM problem and the priority order is derived using AHP. The top prioritised drivers are social sphere (D14), economy (D2) and internet coverage and I.T. Facilities (D4). The prioritised drivers need to be focused on to enhance the feasibility of adoption of cloud manufacturing and Industry 4.0.

TABLE 3.7

Priorities and Ranking of Drivers

Sr. No.	Drivers	Weights	Rank
1	Social sphere (D14)	0.121765	1
2	Economy (D2)	0.108677	2
3	Internet coverage and IT facilities (D4)	0.078465	3
4	Scalability (D7)	0.078465	4
5	Top management support (D16)	0.078465	5
6	Resource sharing and collaboration (D8)	0.066543	6
7	Complexity (D20)	0.050562	7
8	Firm size (D17)	0.04786	8
9	Information sharing (D13)	0.045302	9
10	Knowledge management systems (D3)	0.04288	10
11	Agility (D6)	0.038419	11
12	Relative advantage (D19)	0.038419	12
13	Enterprise architecture (D10)	0.034422	13
14	Technological readiness (D18)	0.032582	14
15	Enterprise integration (D11)	0.029192	15
16	Cloud computing (D1)	0.026155	16
17	Artificial Intelligence (D5)	0.020996	17
18	Smart factory (D12)	0.020996	18
19	Industrial internet of things (D9)	0.020466	19
20	Cyber-physical systems (D15)	0.019372	20

REFERENCES

Carvalho, Núbia, Omar Chaim, Edson Cazarini, and Mateus Gerolamo. "Manufacturing in the Fourth Industrial Revolution: A Positive Prospect in Sustainable Manufacturing." *Procedia Manufacturing* 21 (2018): 671–678.

Castelo-Branco, Isabel, Frederico Cruz-Jesus, and Tiago Oliveira. "Assessing Industry 4.0 Readiness in Manufacturing: Evidence for the European Union." *Computers in Industry* 107 (2019): 22–32.

Ceruti, Alessandro, Pier Marzocca, Alfredo Liverani, and Cees Bil. "Maintenance in Aeronautics in an Industry 4.0 Context: The Role of Augmented Reality and Additive Manufacturing." *Journal of Computational Design and Engineering* 6, no. 4 (2019): 516–526.

Kamble, Sachin S, Angappa Gunasekaran, and Rohit Sharma. "Analysis of the Driving and Dependence Power of Barriers to Adopt Industry 4.0 in Indian Manufacturing Industry." *Computers in Industry* 101 (2018): 107–119.

Lee, Dong Myung, and Paul R Drake. "A Portfolio Model for Component Purchasing Strategy and the Case Study of Two South Korean Elevator Manufacturers." *International Journal of Production Research* 48, no. 22 (2010): 6651–6682.

Liu, Yongkui, and Xun Xu. "Industry 4.0 and Cloud Manufacturing: A Comparative Analysis." *Journal of Manufacturing Science and Engineering* 139, no. 3 (2017): 034701.

Lu, Yang. "Industry 4.0: A Survey on Technologies, Applications and Open Research Issues." *Journal of Industrial Information Integration* 6 (2017): 1–10.

Oliveira, Tiago, Manoj Thomas, and Mariana Espadanal. "Assessing the Determinants of Cloud Computing Adoption: An Analysis of the Manufacturing and Services Sectors." *Information & Management* 51, no. 5 (2014): 497–510.

Roblek, Vasja, Maja Meško, and Alojz Krapež. "A Complex View of Industry 4.0." *Sage Open* 6, no. 2 (2016): 2158244016653987.

Shankar, K, P Kumar, and Devika Kannan. "Analyzing the Drivers of Advanced Sustainable Manufacturing System Using AHP Approach." *Sustainability* 8, no. 8 (2016): 824.

Stock, Tim, Michael Obenaus, Sascha Kunz, and Holger Kohl. "Industry 4.0 as Enabler for a Sustainable Development: A Qualitative Assessment of Its Ecological and Social Potential." *Process Safety and Environmental Protection* 118 (2018): 254–267.

Türkeş, Mirela Cătălina, Ionica Oncioiu, Hassan Danial Aslam, Andreea Marin-Pantelescu, Dan Ioan Topor, and Sorinel Căpuşneanu. "Drivers and Barriers in Using Industry 4.0: A Perspective of Smes in Romania." *Processes* 7, no. 3 (2019): 153.

Vaidya, Saurabh, Prashant Ambad, and Santosh Bhosle. "Industry 4.0–a Glimpse." *Procedia Manufacturing* 20, no. 1 (2018): 233–238.

Vinodh, Shivraman, KR Shivraman, and S Viswesh. "Ahp-Based Lean Concept Selection in a Manufacturing Organization." *Journal of Manufacturing Technology Management* 23, no. 1 (2011): 124–136.

Wang, Xi Vincent, Lihui Wang, Abdullah Mohammed, and Mohammad Givehchi. "Ubiquitous Manufacturing System Based on Cloud: A Robotics Application." *Robotics and Computer-Integrated Manufacturing* 45 (2017): 116–125.

Wu, Dazhong, Matthew J Greer, David W Rosen, and Dirk Schaefer. "Cloud Manufacturing: Drivers, Current Status, and Future Trends." *Paper Presented at the ASME 2013 International Manufacturing Science and Engineering Conference collocated with the 41st North American Manufacturing Research Conference*, Madison, WI, 2013.

Zhang, Lin, Yongliang Luo, Fei Tao, Bo Hu Li, Lei Ren, Xuesong Zhang, Hua Guo, Ying Cheng, Anrui Hu, and Yongkui Liu. "Cloud Manufacturing: A New Manufacturing Paradigm." *Enterprise Information Systems* 8, no. 2 (2014): 167–187.

4 Decision-Making to Achieve Sustainability in Factories

4.1 ARTIFICIAL INTELLIGENCE (A.I.) AND INDUSTRY 4.0

Niraj Kumar, Ashish Das, Lokesh Singh, Padmaja Tripathy, and K. Jayakrishna

A.I.-driven automation has yet to have a quantitatively major impact on productivity growth. In the present day, industries are facing new challenges in terms of market demand and competition. They need a radical change, known as Industry 4.0: integration of A.I. with recent emerging technologies, such as the Industrial Internet of Things (I.I.o.T.), Big Data analytics and cloud computing, that will enable operation of industries in a flexible, efficient and green way.

As A.I. emerges from science fiction to become the frontier of world-changing technologies, there is an urgent need for systematic development and implementation of A.I. to see its real impact in the next generation of industrial systems, namely Industry 4.0.

4.1.1 ELEMENTS IN ARTIFICIAL INTELLIGENCE: ABCDE

The key elements in artificial intelligence are classified by 'ABCDE'.

A: Analytics technology
B: Big Data technology
C: Cloud or cybertechnology
D: Domain know-how
E: Evidence

4.1.2 CHALLENGES OF ARTIFICIAL INTELLIGENCE

1. Machine-to-machine interaction
2. Data quality
3. Cybersecurity

4.1.2.1 Introduction to A.I.

Artificial intelligence (A.I.) is a term or approach through which machines, computers and robots are programmed to think as cleverly as a human can think and make decisions while problem-solving. The intelligence of any system is defined as if the system behaves as intelligently as a human or it behaves in the best possible manner. A.I. is the ability of computer systems, machines or robots to perform intelligent tasks like human beings. What does 'intelligent tasks' include? Learning and acting, reasoning and planning, adopting a new environment, the ability to interact with the real world, understanding natural language, image recognition, robotics, etc. A.I. has achieved success in doing some of the tasks efficiently, like facial recognition, medical diagnosis, driverless cars, planning and scheduling, speech recognition, chess playing at the Grandmaster level, etc. But in some cases, A.I. still needs much improvement, like learning natural language, surfing the web, etc. The A.I. concept comes with two major goals: a) developing an expert system, and b) implementing human intelligence in machines.

4.1.2.2 History of A.I.

John McCarthy first brought A.I. into focus during his first conference in the year 1956. The A.I. concept of making machines mimic and think like humans and whether it is possible to make machines that have the same intelligence as humans to think and learn was introduced by the mathematician Alan Turing. Alan Turing was born on June 23, 1912 in London. He is famous for having decrypted the code of the Enigma machine, used by Nazi Germany to communicate.

Alan Turing performed some tests and raised some questions about whether machines can think. After a long chain of testing, Turing hoped that it would be possible to design such machines that can respond and have memories just like humans. Later this was called the Turing Test. The purpose behind the test was that the machine (computer) could be called intelligent, if a machine (A) and a person (B) exchanged information through natural language and there was also a second person (C), also called the interrogator (Figure 4.1), the Turing test result says that if the second person (C) cannot reliably differentiate between the first person (B) and the machine (A), then the computer is to be considered intelligent (artificial).

Some important features of Turning's test are:

i) It helps to set an objective notion of intelligence i.e., the response of a living intelligence to a particular set of questions. While determining the standard for intelligence, this test avoids any objection to its true nature.
ii) It avoids any confusion about if the computer uses appropriate internal process or whether or not the machine is actually aware of its actions.
iii) It gives an unbiased result by forcing the interrogator to concentrate only on the answers to questions.

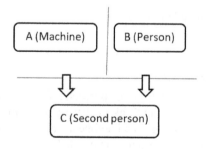

FIGURE 4.1   Turing Test [10].

4.1.2.3 Explanation of Artificial Intelligence

Artificial intelligence is a concept of computer science that conveys the strengths of machines to learn, understand and behave the same as humans. A.I. has been growing rapidly in the past ten years. From a driverless car to Google Search, Apple's Siri and many more that come under the A.I. heading, we have been using them frequently in our daily life and might not have even noticed. A.I. is going to be the biggest breakthrough in the coming years. The backbone of A.I. is machine learning. The term artificial intelligence is self-explanatory, in that it is a technology through which we want to create machines with smart algorithms or sets of instructions to learn based on its knowledge and solve some difficult tasks.

The concept of artificial intelligence draws from many areas:

Computer science, psychology, biology, economics, mathematics, philosophy, linguistics, etc.

The need for artificial intelligence in every sector is increasing exponentially. Computer scientists are predicting that in the coming 5–6 years, 85 per cent of customer interactions will be managed without humans.

'Machines will be capable, within twenty years, of doing any work what man can do.' – Herbert A. Simon (CMU)

'Within a generation … the problem of creating "artificial intelligence" will substantially be solved' – Marvin Minsky (MIT)

4.1.2.4 Typical A.I. Problems

As discussed in previous sections, intelligent systems or intelligent agents of A.I. systems need to do both 'ordinary' and 'Expert' tasks, as shown in Table 4.1.

A.I. has achieved a successful level in solving problems in the area of expert tasks but A.I. has not had the same amount of success in solving the ordinary tasks.

For example, it is found that high-level tasks like proving theorems, playing chess and medical diagnoses are easier to mechanise, but it has been found very hard to mechanise ordinary tasks like walking around without running into things, catching prey and ignoring predators, explaining complex sensory information, etc.

i) A.I. learning and its responses are based on data. So, if the data is inaccurate then it is difficult for an A.I. system to give the correct result.

TABLE 4.1
A.I. System Task [11]

Ordinary Tasks	Expert Tasks
• Planning route, activity	• Medical diagnosis
• Recognising people (vision), objects	• Mathematical problem solving
• Communicating	• Robotics
• Navigation	• Game planning, etc.

ii) Massive unemployment is one of the biggest challenges of the A.I. system, as in many manufacturing industries, robots had already replaced many workers from assembly lines, driverless cars are replacing human drivers and will eliminate them once they are completely implemented.

iii) The security of the data used in A.I.; what if a driverless car is hacked, security fraud, reprogramming drones for criminal uses, misdiagnosing medical patients, identity fraud ... the list goes on and on.

4.1.2.5   Advantages and Disadvantages of A.I.

A.I. is the intelligence of the machines. In recent years, the discussion of the impact of the A.I. on human lives has gained momentum. Initially, A.I. was created to make human lives easier but recently, the ongoing debate is to decide whether A.I. is for or against the future of human existence.

4.1.2.5.1   Advantages

i) **Error reduction.** To achieve accuracy with a high degree of precision by reduction of errors with the help of artificial intelligence. Exploration of space is one of the many studies. Robots with A.I. data and information are sent to explore space with metallic bodies to be more resistive, error-free and with greater ability to explore space and unsympathetic environments. A.I. robots are designed and programmed in such a way that in unsympathetic environments they don't break down or change.

ii) **Difficult exploration.** A.I. robots are programmed in such a way that their performance compared to humans improved, carrying out more laborious and complex work with more responsibility. These robots are sent to do mining, and other fuel exploration. Their intelligence is used more in exploring the ocean floor and places where human reach is difficult.

iii) **Daily application.** A.I. systems like Siri and Cortana have become very common in our daily lives which helps us in many ways with their intelligence. G.P.S. in our smartphones is another example of how we use A.I. when we go on trips and drives. A.I. systems also predict what we wish to type and take some corrective actions against incorrect spelling (machine intelligence). Whenever we post any picture on social media, A.I. algorithms identify the person's face and tags that person. A.I. is currently used in almost every sector, from managing data in banking institutions to financial institutions.

iv) **Repetitive jobs.** Machine intelligence helps in doing repetitious, monotonous work. A.I. systems think faster than the human mind and act faster for multi-tasking. Sometimes hazardous tasks can be carried out with the help of machine intelligence with adjustable speed and time. Computer gaming is one of the wide fields that has used A.I.: machine intelligence plans the game movement according to the movement of the opponent.

v) **Medical applications.** The medical field has wide applications for A.I. Medical surgeons are trained with A.I. robots (surgical) and A.I. simulators. A.I. finds vast application in detecting and monitoring neurological disorders as it can simulate brain functions. Radiosurgery is one of the big uses of A.I. in the medical field. Radiosurgery is very helpful in operating on tumours without damaging the surrounding tissues.

vi) **No breaks**. Machines are programmed in such a way that they can work for long work cycles and with uniform efficiency.

Good reliability, solving complex problems, sometimes preventing data from getting lost and cost-effectiveness are the advantages of A.I. system.

4.1.2.5.2   Disadvantages
  i) **High cost.** Artificial intelligence needs complex machines that require high-cost investment in repair, maintenance and upgrading of software programs to get smarter machines over time.
  ii) **No replicating humans.** The argument continues to date over whether human intelligence is replaced or not. As they lack emotional and social values, machines can only perform according to their programmes and cannot make their own decision on what is correct and incorrect, which sometimes leads to breakdown and is considered the biggest disadvantage.
  iii) **No improvement with experience.** Machines experience wear and tear with time, and unlike humans, A.I. cannot improve in their experience with time as they are programmed.
  iv) **Unemployment.** Introducing A.I. in the future on a large scale is going to be a major issue for unemployment because of so much dependence by humans on machines may lose sense of living.

4.1.2.6   A.I. Models
There are several models that work based on A.I.:

  i) Support vector machine (S.V.M.)
  ii) Artificial neural network (A.N.N.)
  iii) Decision tree
  iv) Fuzzy algorithm
  v) Ant colony algorithm
  vi) Particle swarm algorithm
  vii) Genetic algorithm
  viii) Immune algorithm.

We will discuss some of the above models:

i) *Artificial neural network.* An A.N.N. is an A.I. information processing model based on a biological nervous system, the same as how human brains process information. A.N.N. has many highly interlinked processing elements (neurons) that process information to answer specific problems, as shown in Figure 4.2. Pattern recognition and detecting trends that are difficult to solve for either a human or a computer are one of the applications of A.N.N. through a learning process that is the same way learning in a biological system occurs. In a biological system, learning occurs through adjustment to the synaptic linkage that exists among the neurons. By using this algorithm,

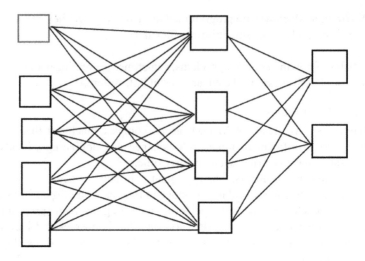

FIGURE 4.2 Artificial neural network [12].

machines identify the problems and help in solving problems just as human brains do.

A.N.N. are programmed in such a way that they think like an 'expert' on given information for problem-solving.

Support vector machine (S.V.M.). S.V.M. is a technique known for classification. S.V.M. is an A.I. model in which a hyperplane is generated that differentiates between the data points of two classes. There are many possible ways to generate hyperplanes but our aim is to get a plane that has the highest margin from support vectors from the two classes. Hyperplanes are conclusion boundaries that help in categorising the data points. If data points are of two types, then the hyperplane generated is simply a line. If the data points are of three types, then the hyperplane generated is a 2D plane, and if there are more than three types of data points, then it is difficult to image the hyperplane, as shown in Figure 4.3.

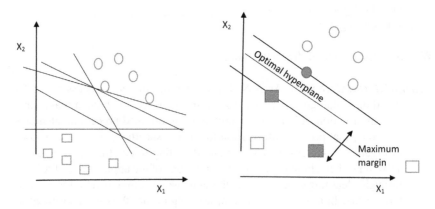

FIGURE 4.3 Possible hyperplanes [9].

Application of S.V.M.:

i) Facial recognition
ii) Hand-writing recognition
iii) Classification of images
iv) Disease diagnosis, etc.

4.1.2.7 Application of A.I.

Willingly or unwillingly, we are all probably using A.I. in our daily life. A.I. has been designed with many algorithms that help the A.I. system to find the expected response.

i) Machine-learning platform
ii) A.I. optimised hardware
iii) Voice recognition
iv) Deep learning
v) Robotics and automation
vi) Decision management
vii) Adaptive manufacturing
viii) Virtual agents, etc.

4.1.2.8 Image Processing through Artificial Intelligence

Images contain a number of pieces of information along with them, and are a great medium to gain and exchange information. Sometimes images carry some faults with them. So, to get some usable information from images, the images must be processed. Image processing is greatly applicable in medical research, aerospace, military, urban planning and many more areas.

Computers and machines treat images as a two-dimensional function. Image processing has three stages: scanning, sampling and quantisation. In the scanning process, the whole image is being traversed; in sampling, the smallest pixels are discretised, and photo-electric sensors are used during the sampling process to obtain the grey values of each pixel. In the quantisation process, with the help of an analogue-to-digital converter, the grey values are converted into discrete values. After the completion of the three above processes, image processing can be studied by different technologies and widely used in different fields.

Image segmentation is one of the technologies which is used in image processing. In this technology, image edges and image regions are studied to obtain meaningful information. To study the edges and regions of images, many methods are used, such as grey threshold segmentation, clustering edge detection and region growth (the concept behind using these methods is that images that are undergoing processing must have obvious grey level changes compared with the background). This concept does not work in the case of complex images like C.T. images, sensor images, remote sensing images and so on. To study such complex images, artificial intelligence along with image segmentation is implemented. Ant colony is an artificial intelligence algorithm combined with image segmentation which gives a good image segmentation effect.

Ant colony: Ants group together and transfer information between them in the process of searching for food. The communication network is possible because ants release a substance (pheromones) during their journey, which helps them in finding the best path for searching for food, and according to this mechanism, a bionic intelligence algorithm has been designed. Ants follow the path with a higher concentration (probability) of pheromones and communicate. In the same way, the ant colony algorithm is a probabilistic method, used to find computational problems which can be reduced to finding the best paths through graphs.

It is found that a larger number of ants searching for food can be considered as positive feedback phenomenon, and the higher the concentration of hormones on a path, the higher the probability of ants choosing that path.

Characteristics of the ant colony algorithm:

i) A.C.A. is based on population, which follows a parallel search.
ii) Every agent (ant) only uses its local information and cannot use global information directly.
iii) Ants can communicate indirectly by changing the environment variable.
iv) A.C.A. can explore several paths at the same time.

Image segmentation has a combination optimisation problem which is later solved by the application of the A.I. algorithm ant colony. Later, the problem of a local solution being used prematurely and stopping the search for the optimal path in the ant colony can be solved by introducing a fish crowding function. The improved ant colony algorithm gives the best improved result on the image segmentation effect. Optimal path search accuracy is improved by an improved ant colony algorithm through simulation.

4.1.2.9 Artificial Intelligence in the Clothing Industry

A.N.N. is an A.I. model which is widely used in the garment industry in the following fields:

i) Mechanical properties prediction
ii) Grading and classification
iii) Fault analysis and identification
iv) Controlling the processes and monitoring
v) Supply chain management.

4.1.2.10 Impact of A.I. on Some Other Industries

i) **Transportation.** With the help of A.I. software, vehicles can park themselves, use adaptive control on long roads, control themselves at traffic signals, convey signals to drivers about obstacles and blind objects during lane changes, etc.
ii) **Robotics (home-based).** Earlier robots used in homes were mostly for cleaning purposes, but after the introduction of sensor-based perceptions, machine learning, physical movement improvements and advancement in speech recognition helps with the use of robots in wider applications.

iii) **Entertainment.** Generation of 3D scenes in movies, music and stage performances, etc.

iv) **Public safety.** Smart watches are the best example of how A.I. is used for safety purposes by sending the exact location of where we are. Drones for surveillance, high security cameras and A.I. can also help in protection from financial fraud, cybersecurity, credit card fraud, etc.

v) **Healthcare.** The medical field uses A.I. mostly and is helping in many ways to improve health and providing quality to millions. Robots are nowadays used to assist surgeons in the operating theatre. The da Vinci robot (surgical) has been used more than half a million times for operations.

vi) **Machine translation technology.** Through this technology, one natural language is converted into another natural language with the help of computer technology e.g., Google Translate.

vii) **Speech recognition.** Speech recognition is one of the advanced uses of A.I. through which speech signals are transferred into corresponding text through a process of interactive human–machine communications, and there are many more sectors around which A.I. is successfully used.

4.1.3.1 Industry 4.0

The idea of Industry 4.0 (I.4.0) comes from German manufacturing. After German industries, Industry 4.0 has been widely accepted by other countries from the European Union, India, China and other Asian countries. Industry 4.0 introduces the fourth industrial revolution, following mechanisation, electricity and I.T. as shown in Figure 4.4.

Industry 4.0 aims in future for industrial businesses to construct networks around the world to link their equipment, industries and warehouses as cyber-physical systems through connecting and controlling mutual intelligence by sharing data to take actions. The cyber-physical units in the form of smart supply chains, smart storage facilities, smart factories and smart machines help in improving manufacturing through product life-cycle management, supply chains and engineering material usage. These units are called horizontal value chain, with the aim that I.4.0 will use each step in the horizontal value chain with great improvement for industrial processes. Industry 4.0 changes the way of production not only through intelligent machines, but also from intelligent products. Intelligent equipment that act intelligently and also products can be provided with intelligence to be identified and located from the start to end of the manufacturing process. The R.F.I.D. tags attached to products enables them to act smartly and helps in knowing their current status,

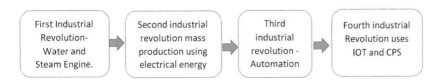

FIGURE 4.4 Industrial Revolution [5].

when they were manufactured, what type of product they are and steps involve to reach their specified location.

This technique allows smart products to know their own history and future and the steps required to convert raw material into finished product. This technique gives the routine in the production process. For example, smart products can give instructions to the conveyor belt, which production line is to be followed and the upcoming production process.

Now we will discuss another element in Industry 4.0, i.e., vertical manufacturing process in the value chain. The idea is to connect a horizontal system with vertical business processes (sales, logistic and finance among others) and associated I.T. systems. It gives smart factories control of the end-to-end management of the entire manufacturing process.

In conclusion, Industry 4.0 needs incorporation of cyber-physical systems (C.P.S.) in manufacturing and logistics with the internet of things (I.o.T.) and other services in manufacturing processes which add new ways to give value and new business models.

4.1.3.2   Defining Industry 4.0

Industry 4.0 is the fourth industrial revolution which brings smart and digital industries with smart production and advanced operational techniques which are not only interconnected with each other and autonomous but can also communicate, analyse and use data intelligently to act. It shows how smart technologies can be introduced to industries, people and assets to give new potential such as artificial intelligence, robotics, nanotechnology, quantum computing, additive manufacturing, advanced materials and the internet of things (I.o.T.).

In short, some definitions say 'making the manufacturing industry fully computerised' and some claims that it is 'making industrial production virtualised'.

4.1.3.3   Why Industry 4.0?

According to earlier concepts in the 1970s and 1980s, with the rise of machines, robots, computers, I.T. is replacing human involvement in manufacturing industries, which may lead to workers losing their jobs and livelihoods. However, Industry 4.0's transformation of manufacturing industries digitally or through combining digital and physical work does not necessarily lead to this problem of downsizing.

Industry 4.0 has some advance technologies. The fast increase of data volumes, cloud storage and rental computing power in recent decades allows analysis of operational data that was not possible earlier. This is now possible with low power consumption and with wide area networks which are now used by many industries accepting it as the future of their manufacturing operations.

i) Industry 4.0 gives advances in analytical capabilities. For good product development, a good analysis is required which increases the quality of the finished product and in some ways, good analysis is required in increasing the efficiency of business operations.

ii) New innovations that make transfer of digital data physically easy have been produced e.g., 3D printing, rapid prototyping.

4.1.3.4   Introduction to the Smart Factory

Industry 4.0 aims to create smart factories in which from supply chain to business models and processes, everything works smartly and beyond our expectations in terms of productivity (NPTEL, n.d.). Smart factories do not only consist of robots and smart machines working together; instead, a smart factory is an entirely linked and flexible system, which can use the flow of data from linked operations and manufacturing systems to acquire and adjust to new demands and brings massive improvements in efficiency and productivity. The smart factory also includes transfer of the linear supply chain into the digital supply chain.

Let's suppose we want to build a smart factory. Then components that we need are:

i) **Smart machines.** Such machines can communicate with other machines, communicate with other smart devices and also communicate with humans.

ii) **Smart manufacturing processes.** Smart manufacturing processes must take care of dynamic changes, automation, real-time changes and be efficient.

iii) **Smart engineering solutions.** Smart design of products, smart development of products, smart planning. Smart machines, smart manufacturing processes and smart engineering solutions together make smart devices and produce output of smart products.

iv) **Information technology.** Smart software application, monitoring, control and smart management processes.

v) **Smart devices.** Connected with themselves, field devices, mobile devices and operating devices.

There are some more components we need when building smart factory like Big Data, smart grid, etc.

Why do we need smart factories?

i) Evolution of technologies
ii) Highly competitive market
iii) High amount of production within minimum timescales
iv) Reduce risk of failure.

Advantages of smart factories:

Reducing costs, increasing efficiency, improving quality, improving predictability and improving safety.

Characteristics of smart factories:

i) **Transparent.** Real-time monitoring, taking required action on time, generating alert messages, real time tracking.

ii) **Proactive.** Predicting quality issues, improving safety, forecasting future outcomes, predicting future challenges.

iii) **Agility.** Flexible, adaptive, self-configuration.

iv) **Connections.** Connected smart devices, smart machines connected with data, connected processes.
v) **Optimisation.** Optimising the task scheduling, use of energy, cost of production, tracking, throughput, reliability.

Use of augmented reality in smart factories (NPTEL, n.d.):

i) Operate instruments remotely
ii) Providing precision
iii) Providing safety especially for radioactive zones.

4.1.3.5 Advantages of Industry 4.0

i) **Improved efficiency.** It has two aspects: a) faster production, b) perfection of the product. Faster production can result from having smart machines that can control the problems and fix them by themselves and smart machine predicting future problems and taking some corrective action against them. Through the implementation of Industry 4.0 in the U.S. and in some Asian countries the production rate has increased by around 25–30%, which is a huge amount. Machines in smart factories have intelligence and are connected to each other, so the working condition of each machine is familiar to the others. This allows them to improve flexibility, speed, productivity and the quality of production processes and products from data analysis-driven improvements that reduce error rates. This is advantageous in industries where demand for high quality is needed, such as medical industries and semiconductors.
ii) **Cheaper production.** In Industry 4.0, machines do everything, which means fewer workers, inspectors and engineers required. In 2016, according to research from Rüssmann et al. (2015), around 4% of the cost (€22 billion) was saved in manufacturing with the help of smart factory systems.
iii) **Agile processes.** Companies are nowadays more responsive to the customer and suppliers' specific needs, with flexible processes which allow for shorter production times and customisation of products.
iv) **Innovations.** With the help of additive manufacturing, more experiments and prototyping can be performed, including meeting customer and suppliers' needs, leading to better manufacturing processes with better results.
v) **Increased productivity.** The production rate gets increased, with a different variety of products with increased quality. Increased productivity is also achieved through improving the supply chain, an improved decision-making system and a distribution system giving better efficiency.

An improved workplace, improved communication, customer satisfaction, increased profits, etc. are some more advantages of Industry 4.0.

4.1.3.6 Disadvantages of Industry 4.0

i) Industry 4.0-based processes need a skilful and educated workforce.
ii) Proactive maintenance of smart machines to reduce production cost is sometimes costly.

iii) Cybersecurity is one of the important aspects.

iv) Stability and reliability are the factors needed for machine-to-machine communications. This demands very accurate and stable timing in the system.

4.1.3.7 Applications

i) Additive Manufacturing in Industry 4.0

Nowadays customisation of products and services is one of the basic demands of customers. To fulfil this demand, Germany brought the concept of Industry 4.0, manufacturing and internet together. Industry 4.0 needs smart factories, smart supply chains, a high degree of automation, flexibility in the processes, etc. (Figure 4.5).

Due to the customisation requirement, along with cyber-physical integration, additive manufacturing becomes one of the important parts in Industry 4.0. A.M. has the ability to create complex objects with advanced attributes i.e., advanced materials. A.M. gives products with increased quality that are used frequently in aerospace, bio-medical and in manufacturing industries. There are some doubts about A.M.'s use during mass production. In future, A.M. may replace conventional manufacturing techniques with its strength of developing complex shaped objects with a high production rate and accuracy. A.M. is not limited only to conventional materials,

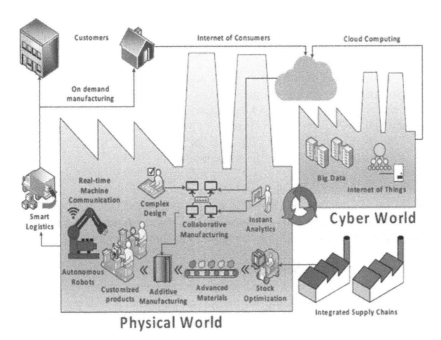

FIGURE 4.5 Smart factory in Industry 4.0. Reprinted from Ugur M. Dilberoglu, Bahar Gharehpapagh, Ulas Yaman, Melik Dolen. 2017. "The Role of Additive Manufacturing in the Era of Industry 4.0." *Procedia Manufacturing* 11: 545–54, Copyright 2017, with permission from Elsevier.

but smart material, metals and social materials are manufactured efficiently and effectively. Due to its cost and manufacturing speed, it may not be preferred in mass production of regular parts. Just because A.M. is able to customise products with complex shapes, sizes, materials and functionality, it will never gain priority over traditional manufacturing processes as shown in Figure 4.3.5. We can conclude that the existence of Industry 4.0 substantially depends on the capabilities of additive manufacturing.

ii) **Application of Industry 4.0 in the Medical Field** (Javaid & Haleem, 2019)

Industry 4.0 provides effective results with interconnection of digital support, I.o.T.s, internal communications, automated machines with sensors, humans to guide them and cybersecurity to secure data. These features of Industry 4.0 help in achieving wide variation in medical and related fields. In medical fields, Industry 4.0 gives a way to produce individual patient-based devices with huge quality outcomes. New technologies help medical organisations to manage them in an up-to-date and competitive state of mind. With the help of sensors and I.o.T., exact information and imaging of complicated cases can be obtained and accurate surgeries can be performed.

- Manufacturing implants suited for individuals can be produced, because Industry 4.0 is flexible with customisation of products. Within less time and lower cost, it gives the best result.
- Customisation of medical parts with the help of Industry 4.0 helps in maintaining a reduced inventory.
- With the help of digital and scanning support, it helps to identify the level, type and reason for any disease.
- Sensors involve in Industry 4.0 decrease the failure rates during surgeries through rapid identification of risk.
- With the customisation of multi-material medical parts, tools and devices with the required strength can be produced.

There are many more applications of Industry 4.0 in the medical field. Industry 4.0 provides smart manufacturing and the use of the exact data at the exact place which helps the medical field to gain a lead towards the new revolution. Through Industry 4.0 in medical fields, wastages can be reduced and human efforts reduced as well.

iii) **Industry 4.0 in Orthopaedics** (Haleem et al. 2019)
- Customisation on a broad scale with the help of smart manufacturing technologies to provide the best outcomes in the orthopaedic field. In less time and with lower cost, it gives the best suited customised implant.
- Orthopaedic technologies, tools and devices are upgraded in the direction of requirements; tools are designed to match specific surgeries.
- Industry 4.0 uses smart devices with the combination of hardware and software to perform controlled surgeries with high accuracy.

- Provides information about leak detection, BP, fluid strain, muscle/tissue strain, strength and temperature of the patient.

iv) **Production of Packaging Films with the Help of Industry 4.0** (Caricato & Grieco 2017)

Through the digital monitoring system and smart techniques, the updated information is available on a daily basis for both manufacturers and customers, and with the help of that information, manufacturing is very efficient and it is also beneficial for end-users making decisions about any product.

v) **Sustainable Manufacturing through Industry 4.0** (Henao-Hernández et al. 2019)

Sustainability and Industry 4.0 are very famous topics nowadays. It is believed that Industry 4.0 brings in profit towards more sustainable processes, such as sustainable manufacturing. In manufacturing processes digital monitoring and process controls are mutual tools with Industry 4.0. As Industry 4.0 is emerging with the concept of intelligent equipment, a more flexible production environment can raise productivity with more efficiency. With the upgrading of equipment and changing technologies and many smart tools, like real-time data in Industry 4.0, it leads to controlled manufacturing through which sustainability is increased.

This study explains that Industry 4.0 has a wide area of variation in technologies to fulfil the sustainable decision-making. To meet the market expectation, fully automated machine tools are the latest trend in the production industries. Industry 4.0 allows sustainability in industries by focusing on low consumption of energy with the help of sensor networks.

vi) **Industry 4.0 in Automobile Industries with Full Automation** (Rüssmann et al. 2015)

Industry 4.0 allows manufacturing industries to adapt automation from a larger scale to a smaller scale, like variations in fabrication methods. Fixed holding devices were used previously, but now autonomous robots can hold and rotate the work as per the needs of robots performing the welding process.

Industry 4.0 allows automobile industries to use a single flexible production line to manufacture different model of cars. Automatic job control techniques can be implemented in future. In addition to helping manufacturers, Industry 4.0 helps vendors to locate their new order automatically to increase the just in time system. This technique will reduce the costs of planning and processes and in many more ways will minimise errors and increase productivity.

4.1.4 Conclusion

- Artificial intelligence is a concept in which machines, robots and computers are programmed in such a way that they can think intelligently like humans and can act and solve problems.

- A.I. can solve expert tasks such as image recognition, driverless vehicles, robotics, playing chess efficiently and also carry out ordinary tasks like navigation, communication, etc.
- Some advantages of A.I. are error reduction, repetitive tasks, no breaks which are the reasons for adoption of A.I. in various fields. Some disadvantages are high implementation cost, unemployment, etc.
- A.N.N., S.V.M., fuzzy logic, ant colony algorithm, etc. are some A.I. models which are being used worldwide in different industries in solving problems effectively and efficiently.
- Applications of A.I. in various fields include medical diagnosis, image recognition, speech recognition, robotics and other industries are increasing rapidly.
- Industry 4.0 is the German industries' concept, which brings manufacturing and internet together and is considered to be a fourth revolution. The success of Industry 4.0 allows countries like the U.S., China and other Asian countries to adopt it.
- Industry 4.0 allows industries to create an intelligent network along with a value chain to control machines autonomously, and to design such a network, six basic principles of Industry 4.0 have to be followed.
- Advantages of Industry 4.0 include improved efficiency, cheaper production, agile process, innovations, increased productivity, improved workplace, improved communication, customer satisfaction, increased profits, etc.
- Disadvantages of Industry 4.0 are the need for an educated and skilled workforce, cybersecurity, etc.
- Applications of Industry 4.0 have much room to expand, due to its innovative characteristics like in the medical field, production industries, to expand automation, etc.

BIBLIOGRAPHY

Caricato, Pierpaolo, and Antonio Grieco. 2017. "An Application of Industry 4.0 to the Production of Packaging Films." *Procedia Manufacturing*, 11: 949–56. doi:10.1016/j.promfg.2017.07.199.
Dilberoglu, Ugur M., Bahar Gharehpapagh, Ulas Yaman, and Melik Dolen. 2017. "The Role of Additive Manufacturing in the Era of Industry 4.0." *Procedia Manufacturing* 11: 545–54. doi:10.1016/j.promfg.2017.07.148.
Haleem, Abid, Mohd Javaid, and Raju Vaishya. 2019. "Industry 4.0 and Its Applications in Orthopaedics." *Journal of Clinical Orthopaedics and Trauma* 10 (3): 615–16. doi:10.1016/j.jcot.2018.09.015.
Henao-Hernández, Iván, Elyn L. Solano-Charris, Andrés Muñoz-Villamizar, Javier Santos, and Rafael Henríquez-Machado. 2019. "Control and Monitoring for Sustainable Manufacturing in the Industry 4.0: A Literature Review." *IFAC-PapersOnLine* 52 (10): 195–200. doi:10.1016/j.ifacol.2019.10.022.
Industry 4.0 and IOT: Joachim von Heimburg. https://innovationmanagement.se/2016/12/29/industry-4-0-and-the-internet-of-things-iot/.
Javaid, Mohd, and Abid Haleem. 2019. "Industry 4.0 Applications in Medical Field: A Brief Review." *Current Medicine Research and Practice* 9 (3): 102–9. doi:10.1016/j.cmrp.2019.04.001.

NPTEL. n.d. *Video Lecture on Smart Industries.*
Rüssmann, Michael, Markus Lorenz, Philipp Gerbert, Manuela Waldner, Jan Justus, Pascal Engel, and Michael Harnisch. 2015. "Industry 4.0: The Future of Productivity and Growth in Manufacturing Industries." The Boston Consulting Group. 2015.doi:10.1007 /s12599-014-0334-4.
SVM, Nandini Jella et al. https://www.ques10.com/p/41200/support-vector-machine-1/.
Wikipedia: https://en.wikipedia.org/wiki/Turing_test.
Wikipedia: https://en.wikipedia.org/wiki/Artificial_intelligence.
Wikipedia:https://en.wikipedia.org/wiki/Artificial_neural_network.

4.2 ROLE OF MACHINE LEARNING IN INDUSTRY 4.0

Shambhu Kumar Manjhi, Ashish Das, Shashi Bhusan Prasad, Lokesh Singh, Padmaja Tripathy, and K. Jayakrishna

4.2.1 INTRODUCTION

Computer scientists are very much excited about artificial intelligence and machine learning which will let machines be as intellectually capable as human beings in terms of decision-making. It has ignited their imagination, and they are striving to make an intelligent system which will shape the future of the industry. Machine learning is an important part of this system [1].

4.2.2 HISTORY OF MACHINE LEARNING

In 1950 Alan Turing created a test called the 'Turing test'. It became the check of the machine's capacity to think like a human. To pass the test, it has to convince someone that it is also human.

In 1952, Arthur Samuels wrote a program which could play the game of checkers. As it played more, it improved its winning strategies and then it integrated those into its program. He coined the term 'system of learning'. According to him, system mastering is a 'Field of observing that gives computers the potential to research without being explicitly programmed'.

In 1957 Frank Rosenblatt proposed a simple neural network called 'perceptron' which could mimic the thought processes of the human brain.

In 1967 the 'nearest neighbour algorithm' was written which could be used for basic pattern recognition.

In 1982 John Hopfield created a neural network which had bidirectional lines. It worked similarly to how neurons work in humans.

In 1997 IBM computer Deep Blue defeated world chess champion Garry Kasparov.

In 2006 Geoffrey Hinton explained new algorithms which could detect objects and text in images and videos. He coined the term 'deep learning'.

In 2011 IBM's Watson beat a human in a game of *Jeopardy!*.

In 2012 Jeff Dean of Google created the Google Brain which uses a deep neural network for pattern detection in images and videos. It was used for image detection in YouTube videos.

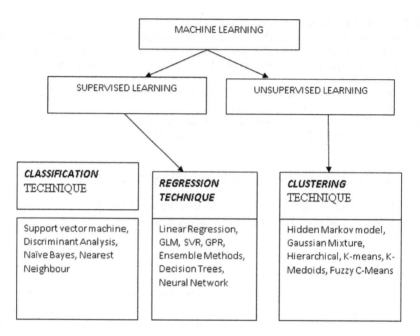

FIGURE 4.6   Classification of machine learning [1].

In 2014 Facebook developed the DeepFace algorithm which can identify people's face with the same precision as that of the human.

In 2016 DeepMind researchers developed an algorithm AlphaGo which could play the Chinese board game Go. It was considered to be the world's most complex board game. It managed to win all the games in a Go competition.

4.2.3   MACHINE LEARNING

Machine learning is a utility of artificial intelligence. The machine can analyse and enhance their experience without being explicitly programmed. Machine learning specialises in the improvement of algorithms that could get right of entry to data and use it to learn for themselves.

The process of learning begins with the observation of data, provision of instruction and by direct experience. Machine learning models are then built using data and algorithms which can be used to make better decisions in future [1].

4.2.4   BROAD CLASSIFICATION OF MACHINE LEARNING

Machine learning can be classified as supervised and unsupervised learning, as shown in Figure 4.6.

4.2.4.1   Supervised Learning

In supervised learning, we have information which can be enter variable (x) and output variable (y). We use a model (algorithm) to study the mapping characteristic

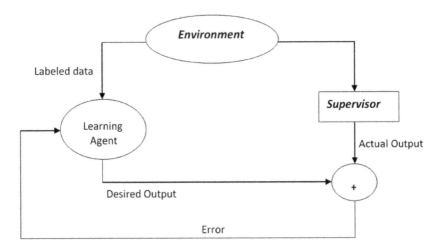

FIGURE 4.7 Process diagram of supervised learning [2].

from the input to the output. In this model, we have a supervisor and a learning agent as shown in Figure 4.7. Both the supervisor and the learning agent interact with the environment i.e., they are given the input data (x) [2]. The supervisor can be thought of as a teacher who has the desired output set. When a new input (x) is fed in, the algorithm provides the actual output which is compared with the desired output by the supervisor and is corrected. This is an iterative manner which stops when an acceptable level of performance is complete [2].

Supervised learning uses classification and regression models.

4.2.4.1.1 Classification Technique
This is used to identify the class the object belongs to by using one or more independent variables. It is used to predict a discrete response. For example, to determine whether an email is genuine or spam, to recognise letters and numbers, etc. [2]. Figure 4.8 shows the classification of customers applying for credit card according to risk parameters. Some algorithms to perform classification are support vector machines, K-nearest neighbour, neural network, etc.

4.2.4.1.2 Regression Technique
It is a predictive modelling method which offers counting among biased and unbiased variables and is also used to predict continuous responses. For example, change in temperature of a machine part, vibration level in a machine part, forecasting, time series modelling, etc. Some algorithms that perform regression are decision tree, neural network and fuzzy learning [3].

4.2.4.2 Unsupervised Learning
In unsupervised learning, we have only input data (X) and no predefined analogous output data is provided. In this model, we have only the learning agent that is fed with data and here the supervisor is absent, as shown in Figure 4.9. The learning

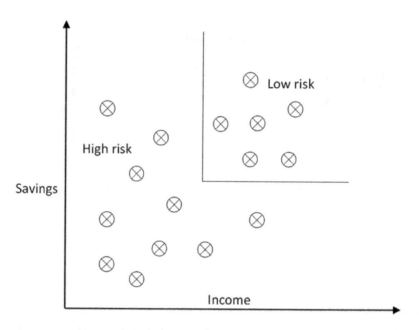

FIGURE 4.8 Classification of the person applying for a credit card [2].

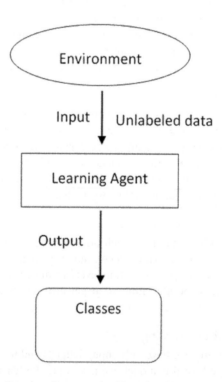

FIGURE 4.9 Process diagram of unsupervised learning [4].

agent itself finds the pattern from the data. Unsupervised learning is more complex, but it can also solve more complex problems [4].

4.2.4.2.1   Clustering

The usual unsupervised studying method is clustering. Clustering is the approach to discovering similar groups of information in an information set. The objects of each group are comparatively more similar to objects of that group than those of other groups. The aim is to segregate the population with similar traits, as shown in Figure 4.10 and Figure 4.11. For example, mobile network companies can use clustering methods to identify the best locations for their cell towers to optimise signal reception for the customer by estimating the number of clusters of people relying on their network in a given area [4].

4.2.5   METHODS OF LEARNING

Depending upon the method of learning, task machine learning can be classified into several categories.

4.2.5.1   Concept Learning

Human beings learn by acquiring a concept from their past experiences. For example, humans can identify an animal, say, an elephant, based on some specific sets of features, like it has a trunk, big ears, etc. This feature which differentiates an elephant from other animals is called a concept. Similarly, a machine also learns a

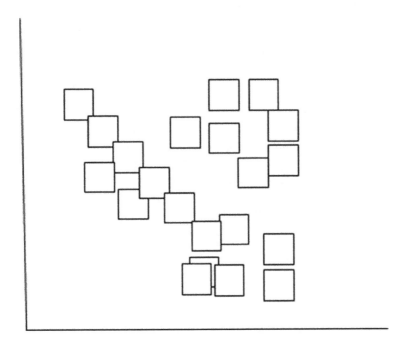

FIGURE 4.10   Before clustering [4].

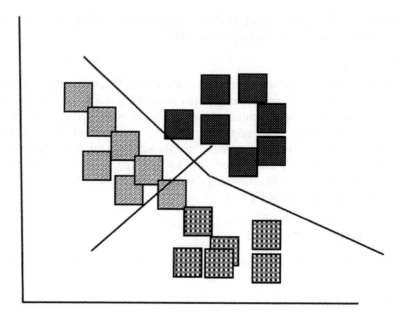

FIGURE 4.11   After clustering [4].

concept by making different sets of attributes and then compares whether the attributes of an object belong to a specific set [5].

For concept learning, an algorithm requires:

- Training data (data from our past experiences)
- Target concept (hypotheses to identify data object)
- Actual data object (a new data object on which testing is done).

4.2.5.2   Decision Tree Learning

A decision tree is a predictive learning model. In decision tree learning, we split the dataset using an algorithmic approach based on various conditions. It is the most widely used method for supervised learning. The goal of this model is to predict the value of a target variable by applying simple decision rules inferred from the data features [5].

Conditions are implied in the form of if-then-else statements. As the level of the tree increases, the condition gets more complex and the model gets better.

The structure of a decision tree is an upside-down tree with roots at the top, and its branches increase as we go down, as shown in Figure 4.12. The decision tree consists of nodes, edges and leaves. The node is the place where we place an attribute and ask a question. The edges answer this question and the output is represented by the leaves [3].

Each node acts as a test case and the edges give the possible answers. This is a recursive process and is repeated for every subtree.

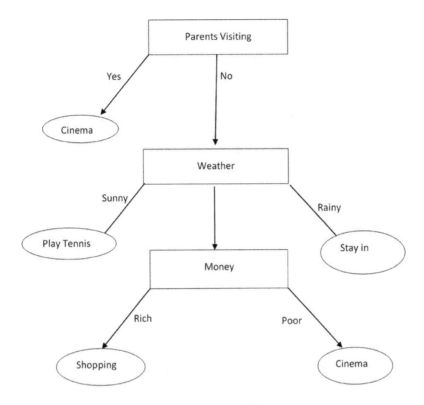

FIGURE 4.12 An example of a decision tree model [5].

4.2.6 PERCEPTRON LEARNING

A perceptron is the simplest type of artificial neural network. A perceptron learning algorithm is a computational model. A perceptron accepts some training data and assigns some weight to it. We combine the weights and training data in a linear equation called activation.

Activation = sum{Weight (i) * $X(i)$ + bias

The output returned by this algorithm is either 0 or 1. If activation is greater than some threshold value, the output is 1, and if the activation is less than that threshold value, the output is 0 [6].

4.2.6.1 Bayesian Learning

Bayesian learning takes into account the probabilistic function in the algorithm. Suppose you are given a new coin with a head and a tail on its two sides [1]. You are asked to determine the fairness of the coin. To determine this, you have done ten coin-flips, and suppose you get six heads and four tails. With this observation, you cannot be sure that the coin is unbiased. You decided to increase the number of observations and you get the results shown in Table 4.2.

When you increase the number of trials, it guarantees accuracy. Bayesian learning works on this principle. As we give more data inputs to our algorithm, the machine will update its belief and the confidence level will increase on the results [7].

TABLE 4.2
Valuable Data with the Number of Heads and Their Probability [5]

Number of Coin Flips	Number of Heads	Probability of Observing heads
10	6	0.612
50	29	0.582
100	55	0.556
200	94	0.470
500	245	0.490

4.2.6.2 Reinforcement Learning

The goal of reinforcement learning is to develop a good action for a particular situation. This is done through digital rewards and punishment. When a reinforcement learning algorithm is given an input, it provides an output. The output may be favourable or unfavourable [7]. If the output is favourable, the model is given a positive reward, and if it is unfavourable, it is given a negative reward.

Gradually with time, it filters out the unfavourable outputs, and it improves with time. Figure 4.13 shows the process diagram of reinforcement learning.

4.2.7 Artificial Neural Network and Deep Learning

4.2.7.1 Artificial Neural Network

The artificial neural network combines both biology and arithmetic collectively. The artificial neural network is stimulated by using humans' biological neural network. In our biological nerve cells i.e., neurons have extraordinary components like dendrites, cell frames, axons and synapses. The dendrites are the input factors for any records [8]. The statistics are transferred to the neurons from any sensory point of the body or other dendrites. The mobile frame processes those records and sends outputs in the form of the electrical signal via axons to other neurons or motor output. The synapse is the factor wherein the output of one neuron is transformed so that it can be general with the aid of other neurons [8]. In Figure 4.14, we will see a synthetic neural network. It has a layer of networks: the input, the centre, and the

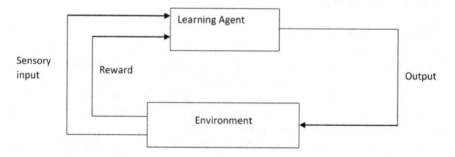

FIGURE 4.13 Process diagram of reinforcement learning [7].

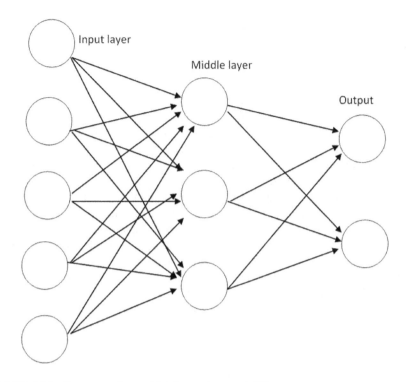

FIGURE 4.14 Structure of an artificial neural network [9].

output layer. Likewise, it can have multiple layers. We can think of every circle as a neuron, so every neuron accepts input techniques and generates an output. When the output generated in the first neuron is passed to the subsequent neuron, a weight gets multiplied to it and a brand-new cost receives conventional by using neuron as input. The input and output can be compared to the dendrites and axons of a biological neural community [9]. The weights expanded may be in comparison to the feature of the synapse.

4.2.7.2 Deep Learning

Deep learning is a specific form of machine learning. It is a learning technique that teaches the machine to make decisions exactly as human beings do [10]. It is possible with the help of algorithms whose structure and function are similar to that of the human brain i.e., an Artificial Neural Network. Deep learning achieves such a high level of accuracy with the help of a large amount of trained data like images, text, sounds and a neural network architecture that contains many layers. It can easily outperform humans [10]. It requires a high degree of computational power and in this regard cloud computing helps in reducing its training time. Deep learning is an extended version of the neural network. Deep learning and neural network both work on perceptron and are made up of multiple layers, but the difference is that we go much deeper in deep learning. That means we use more layers than usual in an artificial neural network to achieve deep neural network, and we call it deep learning [11].

4.2.8   What Can Machine Learning do?

4.2.8.1   Data Mining

Data mining is the process of finding a useful set of data from a large collection of data to identify patterns and relationships among data and allow industries to find future scope. Machine learning has made it easier to extract hidden knowledge in a large volume of data. For example, the data mined using a database of manufacturing and assembly of a turbine rotor can be used to solve a design-related problem. The machine learning algorithm can find the relationship between balance and vibration of the turbine which can be used to improve the tolerance design of parts when producing parts in future [11].

One fine example of data mining is 'The Materials Project' by the Materials Genome Initiative (M.G.I.). In this project, the machine learning algorithm is fed data of various materials whose composition and properties are known. With the help of data mining and learning techniques, it can predict the properties of various new materials before those materials are synthesised and tested.

Industries have stored a huge set of data about customers, business partners and suppliers. These data have hidden useful knowledge in them which can be extracted using machine learning algorithms. This will help the company to optimally manage and allocate their resources to their most important customers and suppliers [12].

4.2.8.2   Quality Management

Machine learning can be useful in product quality inspection, fault detection and defect forecasting.

4.2.8.2.1   Quality Inspection

Using deep learning technology, we can incorporate generic features of a component in a machine learning algorithm like a wide range of textures, surface roughness and various other non-conformities on the surface of various materials. Using neural network architecture and a backpropagation technique, we can automatically inspect for wear, scratches, burrs, holes and other difficult to detect defects. This can reduce human labour and improve the quality of the product [12].

4.2.8.2.2   Fault Diagnosis

Often failures are caused due to undesirable operating conditions, excessive loads, overheating, corrosion and wear. This may lead to an increase in production costs and lower productivity. With the help of smart sensory devices installed over machines at various locations, I.o.T.s and the cloud, we can collect data on operating conditions. These data are fed to the algorithm which can diagnose the root causes of failure [13].

4.2.8.3   Predictive Maintenance

Predictive maintenance is used to predict which machine or component is going to fail and at what time. This will help to develop a good maintenance strategy, prevent major breakdowns, reduce downtime and reduce maintenance cost. Preventive maintenance is possible in Industry 4.0 with the help of different sensors interconnected with I.o.T.s and the cloud. With long- and short-term memory, a recurrent neural network is useful in predicting defect occurrence and estimating the remaining useful lifespans of mechanical systems or components [13].

4.2.8.4 Supply Chain Management

Machine learning can be used in the field of supply chain management to detect inefficiencies and add better control to the process. The areas in which machine learning has an impact on supply chain optimisation are as follows:

4.2.8.4.1 Optimising Supply through Data

The machine learning algorithm can handle an enormous amount of data and more diverse data, so they can improve demand forecasting. Datasets such as historical performance, promotions, media and market modelling are usually used for demand planning. Machine learning also adds less data-specific values such as promotional lift, halo effect lift attributed to a product segment, seasonality and web presence or social media [14].

These variables are usually perceived by human perception, but machine learning can apply them at a better level to improve the accuracy of forecasting and reduce the rate of error.

4.2.8.4.2 Improvement in Shipment Capabilities

The people involved in the supply chain network spend a considerable amount of time and effort in checking the products arriving or departing at the loading and unloading dock. The physical inspection of assets and their maintenance can be reduced by using machine learning's visual and pattern recognition abilities. At the primary level, machine learning can be used to validate the packaging by recognising it and counting the quantities. On a more advanced level, it can be used to identify wear conditions on production line equipment and check vehicles for wear and evidence of accidents [15].

4.2.8.5 Process Planning

Process planning is an excellent area of the utility of machine learning methods.

4.2.8.5.1 Identification and Processing of Machining Features

We can use machine learning to recognise a machining feature from a 3D model. After recognition of machining features, machining instructions such as tool entrance face, drive face and part face can be generated by the use of suitable algorithms and further dealt with to produce N.C. codes. When a new feature cannot be recognised by the learning agent, it may lead to unforeseen feature interactions. The system can be trained to absorb those features using a supervised learning technique. After a large amount of data for machining features is fed to the system, the system can then automatically recognise the machining feature of a new part using a boundary shape. This can be used for process planning of machining features that have shapes similar to memorised ones [15].

4.2.8.5.2 Part Families Machining Cells and Group Technology

Machine learning can contribute to reducing the effort in industries for classification and manual coding of parts, the formation of elemental household and corresponding machining cells. Conceptual clustering has a specific role here in clustering the objects and formation of hierarchies [15].

4.2.8.5.3 Part Grouping in a General Setting

Classification of parts can be done using the identification of geometrical shapes and machining attributes simultaneously. One such classification technique that has been developed is I.P.F.A.C.S. (image processing and fuzzy art-based clustering system). It carries picture processing techniques and changed fuzzy art neural community set of rules. With the help of image data and scalable property, parts can be classified. There can be various other shape recognition approaches [16].

4.2.8.5.4 Part Grouping with the Aim of Machining Cell Formation

Recognition of element families that require similar production and ordering of machines is a crucial part in manufacturing industries. The formation of an element family, a device cell identity, bottleneck machine detection and herbal cluster generation can be carried out with the use of the self-organising neural network. The neural network has the benefit which makes it viable to combine any new component to current machines cells without repeating the entire computational manner. Part grouping with emphasis on features known or recognised using machine learning parts can be grouped by the features they possess such as shape, edges, pockets, slots, steps, holes, etc. Forming of part families is a very difficult and laborious task when, typically, thousands of different features are present. This can be eliminated using machine learning. A parallel distributed processing (P.D.P.) model having three layers of a synthetic neural network has been used in developing a model for classification of features [16].

4.2.8.6 Operation Selection and Planning

Operation planning is used to determine the optimal set of operational parameters that can be used during the operation to get the product that meets the required specification. Researchers have developed a model that with the help of various numerical algorithms can serve as a simulator of the machining process and predict their performance. Engineers can use this prediction to improve their decision. The domain knowledge from machining handbooks, heuristics and databases are fed to the algorithm and the simulator is used to generate training examples. Proper decision points for simulation are selected using designs of experiment concepts. The output of the simulation is compiled into a data file that contains decisions and parameters. The compiled data are then grouped into classes. Each class of data will have separate decision rules for the operation which can be used for consultation among planner for operation planning. Figure 4.15 shows the architecture for operation selection and planning [17].

4.2.8.6.1 Operation Sequence Planning

Process planning functions can use the techniques of machine learning to automate the planning process. The planning domain consists of a set of type of objects,

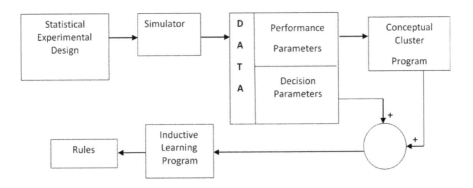

FIGURE 4.15 Operation selection and planning architecture [17].

library of operators and inference rules that act on these objects. Each operation acts as a decision point. Learning acts at each decision point which aims to improve the learning efficiency and improve the quality of plans. Learning modules consists of rules, abstraction mechanism, analytical reasoning ability and learning by experimentation and observation. The quality difference between various plans can be studied and improvement plans can be generated using a search tree. Various planning techniques for operation sequence planning will be playing a promising role in Industry 4.0 [18].

4.2.8.6.2 Process Modelling Monitoring and Control

The conditions of manufacturing processes are subjected to various machining parameters like temperature, force, torque, etc. These parameters are subject to be controlled by various control and monitoring algorithms. Sensor integration for data collection is necessary for this field [18]. For sensor integration, various processing techniques can be used, like Bayesian and Multi-Bayesian estimation, statistical decision theory, fuzzy logic and production rules. Tool condition monitoring, process monitoring and adaptive control issues can be solved using pattern recognition, an artificial neural network and a hybrid artificial intelligence system.

4.2.8.7 Tool Condition Monitoring

Tool condition monitoring is a bit of a difficult task because it requires intelligent sensors which can do:

- Self-calibration
- Signal processing
- Decision-making
- Fusion ability.
- Learning Capability

The unsteady nature of cutting processes like steady-state cutting, chatter and built-up-edge formation can be dealt with by fuzzy pattern recognition. It can also be used to monitor flank wear in tool flanks and the rate of the severity of the wear.

Multiple sensors can also be integrated using an artificial neural network to get better results [19].

4.2.8.8 Process Modelling

In computer integrated manufacturing, reliable process modelling is very important. They are required for optimal parameter design and implementing an adaptive control system. Multi-variable and non-linear estimators like neuro-fuzzy (N.F.) and the artificial neural network system are generally very effective in the process modelling approach [19].

In Figure 4.16, you can observe the input variables like depth of cut (d), current (i) and feed (f) are fed to the neural network. The learning model permits the controller to understand the impact on output variables like power (P), temperature (T), surface roughness (R) and cutting force (F).

4.2.8.8.1 Adaptive Control

In the traditional machining process, the machining parameters like fed rate, depth of penetration, cutting velocity, etc. are usually done before production starts with no provision for in-operation adjustments of these parameters. Due to this, optimal use of the machine in terms of performance parameters is hindered. Incorporation of online adjustment of performance parameters is possible with the help of machine learning. A suitable model using an artificial neural network is made which can map the input variables (feed, depth of penetration and cutting velocity) to get the output variables like cutting force, power, temperature and surface finish). The inverse of this mapping can be used for adaptive control (prediction of the machining parameters that optimise the method pleasurable the constraint on measured or output variable). Figure 4.17 illustrates the difficulty [20].

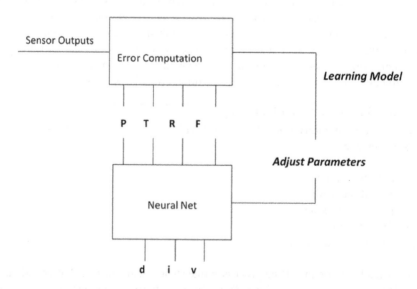

FIGURE 4.16 Learning of process models in machining [19].

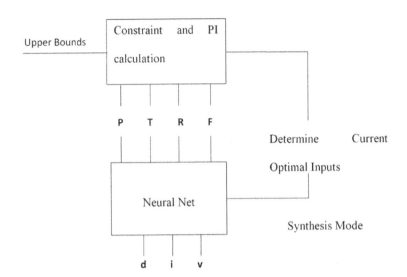

FIGURE 4.17   Synthesis phase of neural network-based adaptive control [20].

This technique is very much useful in the wire E.D.M. process because in the wire E.D.M. process, it is hard to decide optimal cutting parameters for improving reducing performance [20].

4.2.8.8.2   Production Planning and Control

The historical data of the manufacturing system is used for scheduling selections in dynamic surroundings. Machine learning may be used to check machine overall performance with the aid of growing weighted aggregate of heuristic priority rules. Operational rules may be modified using the artificial neural community-based gadget, where operational regulations are weighted characteristics of criteria which includes suggesting technique, price or schedule. Performance measures are represented by quality measures like cost within some past period [21].

At the first set of some training, instances are created beneath diverse masses of WorkCentre and through the operational regulations of this WorkCentre. Thereafter we can provide input as workload and performance measures and artificial neural network provide suitable operational policies. It is a method of multiple-criteria selection-making where criteria weights are a result of simple and automatic schooling method.

Figure 4.18 shows a scheduling answer. Here the variable dispatching scheme is relevant underneath numerous performance measures, situations and events that occur in dynamic operation surroundings. Here machine learning takes a sequence of decisions containing alternative routing, machine and machining according to the dispatching rule based on current system status. These choices are initiated by using the machines earlier than the operation starts. The right selection of selection policies relies upon a fine-tuned status of the present state of the device [22].

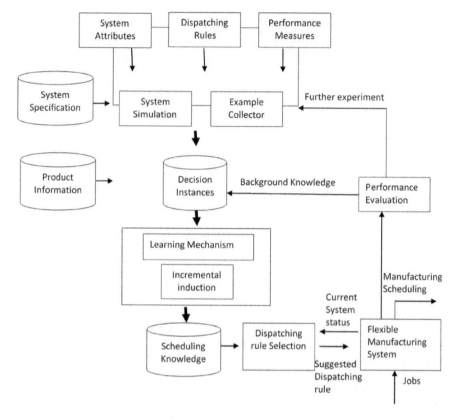

FIGURE 4.18   Architecture of the learning-based dynamic scheduling system [21].

The problems to be solved are:

- Extraction of the scheduling know-how, and
- Search for the good schedule.

The first problem is solved by the decision tree. The second problem is solved by a genetic algorithm.

4.2.8.8.3   Automated Inspection Diagnostics and Fine Manipulation

Major completion in today's industrial era is to reduce the cost and increase the quality and reliability of the manufactured parts. Inspection diagnostics and fine manipulation are therefore a very important function of manufacturing. Therefore, control and supervision of process parameters are important to attain the desired dimensional and floor best. Researchers are working on an expert system using machine learning to build a model for diagnostics [22].

The first step is to build an initial knowledge base using a description primarily based on the mastering method. Using this model generates regulations for the function of the manufacturing system. When the model is applied to a production process, using analogical inference and a similarity-based knowledge technique, it improves the system's diagnostic capabilities.

C.M.M. (co-ordinate measuring machines) or a different role and distance mea-suring device e.g., a laser scanner is used to discover the location and size of an item. The measurement facts received are used for the cause of inspection or opposite engineering. The artificial neural network is used to set up the relationship among measured values and operating parameters. This can be used to correct the error positioning by the machine and improve its performance. The following measures need to be taken in in an integrated way, as shown in Figure 4.19.

• Attainment of in-manner tracking of tactics and workpiece
• Development of dependable technique models applicable to real situations
• Implementation of learning technique for exploration of hidden features, influences, relations and effects and their usage in further manufacturing.

4.2.9 Applications of Machine Learning

4.2.9.1 Manufacturing Industry

Machine learning has revolutionised how we can interact with machines and data. In the manufacturing sector, machine learning can provide a lot of advantages like

• Increase in productivity
• Minimisation of human errors
• Optimisation of production costs
• Repetitive tasks can be done easily.

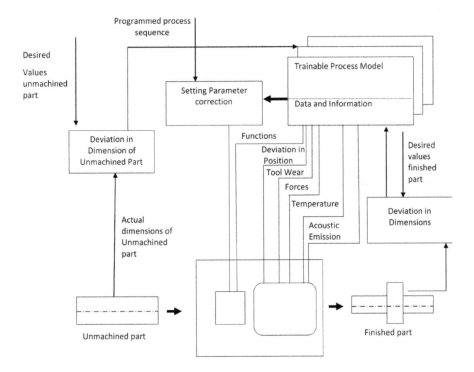

FIGURE 4.19 Quality control loop for zero-defect manufacturing [22].

The more data we feed to the computer, the more the effectiveness of machine learning will increase in observing the trends and making decisions which benefit the organisation. In the manufacturing industry, asset management, supply chain management and inventory management are major areas in which we can implement machine learning.

Industries can adopt predictive maintenance i.e., to determine when a maintenance action needs to be performed, through analytics tools using machine learning. This will lead to a reduction of cost through lower labour cost, less maintenance activity and less wastage.

Improvement in supply chain management through efficient inventory management, reduction of supply chain forecast errors which will lead to better availability of product to the customer and a synchronised pattern of production flow [23].

Automatic guided vehicles (A.G.V.s), cranes and other material handling devices can be made more intelligent in automatic handling of materials such that they could pick up, transport and store the right amount of material, at the right time and in the right place without any human interference. This will lead to the reduction of human labour, effective flow of material, decrease in production time and reduction of production costs.

Consumer-focused manufacturing will be easier as any change in market demand will be captured easily and the company can respond quickly according to the customers' need [24].

4.2.9.2 Finance Sector

Despite many challenges, financial companies take advantage of this technology. The advantages in the financial sector are:

- Operational costs are reduced due to process automation
- User experience is enhanced leading to better productivity and increased revenue
- Better security and compliance.

Some of the machine learning use cases in finance are:

4.2.9.3 Process Automation

Automation with the help of machine learning allows for the reduction of manual work, automation of repetitive tasks and increases in productivity. Accordingly, machine learning empowers organisations to optimise costs, improve customer experience and scale-up administration. Here are some automation use instances of machine learning.

JP Morgan Chase and Co. is a prime global financial administration company and one of the biggest financial establishments in the United States drove a Contract Intelligence (COIN) platform that uses natural language processing, one of the A.I. structures. The company administration takes official documents and distils primary records from them. An annual audit of 12,000 yearly business credit score statements could commonly soak up round 360,000 paintings hours [25].

BNY Mellon co-ordinated automation into their financial banking community. This development is in charge of $300,000 in yearly investment funds and has realised a wide scope of operational upgrades.

Wells Fargo makes use of an A.I.-driven chatbot through the Facebook Messenger application to speak with customers and offer help with passwords and information.

Privatbank is a Ukrainian financial organisation that actualised chatbot workers over its versatile and net platform. Chatbots raised the goals of general patron queries and were accepted to decrease the amount of human help.

4.2.9.4 Security

Security dangers in funds are increasing along with the growing quantity of change, clients and third party mixes. What's more, machine mastering algorithms are fantastic at figuring out frauds.

For instance, banks can utilise this innovation to display many wonderful trade parameters for each report constantly. The calculation analyses each transaction a cardholder makes and analyses if an attempted transfer is usual for that precise purchaser. Such a model spots fraud behaviour with excessive exactness [26].

On the off-chance that the framework spots suspicious conduct, it could ask for additional ID from the client to approve the exchange. Or then again, even square the alternate interior and out, if there's at any rate a 95% likelihood of it being a fake. A machine gaining knowledge of calculations needs only a moment (or maybe seconds) to evaluate a change. The speed prevents fraud in real time, and as a result, prevents any crime.

Money-related checking is another safety use case for machine learning in finance. Information researchers can put together the framework to distinguish an excessive number of micropayments and flag such tax evasion techniques as smurfing.

Machine learning calculations can essentially enhance community safety, as well. Information researchers train a framework to spot and separate cyber-threats, as machine learning is fine in class in breaking down a large range of parameters in real time. Furthermore, this innovation-decision will shock the most advanced cybersecurity community within the closest destiny.

Adyen, Pioneer, PayPal, Stripe and Skrill are some eminent fintech organisations that worked hard on security machine learning.

4.2.9.5 Guaranteeing and Credit Scoring

Machine learning calculations fit well with certain tasks which are so frequent in finance and security. Information researchers learn trends from a huge range of consumer profiles with numerous data entries for every client. A well-trained framework would then be capable of playing out the equivalent approval and credit scoring tasks in real situations. Such scoring vehicles help human representatives' paintings faster and more exactly [27].

Banks and coverage groups have plentiful customer information, and a good way to utilise those records is to put together machine learning models. Then again, they can use datasets produced through substantial telecom or provider groups.

For example, BBVA Bancomer is working together with a voluntary credit-scoring platform Destacame. The bank means to create access for clients for a low record of loan repayment in Latin America. Destacame gets payments data from service organisations utilising open A.P.I.s. Utilising payments conduct, Destacame produces a F.I.C.O. assessment for a client and sends the result to the bank.

4.2.9.6   Robo-Advisors

Robo-advisors are currently normal in the economic area. As of now, there are noteworthy uses of machine learning in the security area. Portfolio control is an internet wealth control management system that makes use of calculations and insights to allot, oversee and streamline customers' assets. Clients enter their present money, associated assets and goals, kingdom, sparing a million bucks by way of the age of 50. A robo-consultant at that factor assigns the prevailing assets crosswise over investment opportunities depending on the risk tolerance and the suitable objectives.

Numerous online insurance administrations use robo-advisors to prescribe customised protection plans to a specific client. Clients pick robo-advisers over other financial consultants because of lower expenses, just as with customised and adjusted proposals.

4.2.9.7   Healthcare Industries

Advances in machine learning and artificial intelligence are transforming our ability to analyse the vast amount of data and to predict outcomes in both bio-medical research and healthcare delivery. Machine learning helps manage the massive amount of data being collected and it can improve clinical decision-making, clinical diagnoses and ultimately, it can affect our healthcare.

It holds so much promise, yet it is not without challenges.

Machine learning algorithms can be trained to arrive at the right diagnosis at expert level performance; for example, in 2016 a group of researchers from Google reported a computer system diagnosing diabetic retinopathy which is a common complication of diabetes by looking at the fundus photograph [28].

4.2.9.8   Cancer Diagnosis

Conventional strategies for identifying and diagnosing cancer growths incorporate computed tomography (C.T.), magnetic resonance imaging (M.R.I.), ultrasonography, and X-ray. Sadly, numerous cancerous growths can't be analysed precisely enough to dependably save lives with these systems. Examination of microarray gene profiles is an option, yet it depends on numerous long stretches of calculation – except if that investigation is machine learning-empowered. Stanford's A.I.-empowered diagnostic algorithm has now been demonstrated to be as successful at recognising potential skin tumours from pictures as a group of 21 board-certified dermatologists. The start-up Enclitic is utilising deep learning how to recognise lung cancer growth nodules in C.T. pictures—and their calculation is 50% more exact than specialist thoracic radiologists functioning as a group.

Tuberculosis (T.B.) is a worldwide issue that truly imperils general well-being. Pathology is a standout amongst the most vital methods for diagnosing T.B. in clinical practice. To affirm T.B. as the diagnosis, finding exceptionally recoloured T.B. bacilli under a magnifying lens is standard. Given the exceptionally small size and several bacilli, it is exacting and demanding work, even for experienced pathologists, and this difficulty frequently prompts low identification rate and false findings. The man-made brainpower of machine learning can help identify corrosive quick recoloured T.B. bacillus.

4.2.9.9 Detection of Haemorrhage

Intracranial haemorrhage (I.C.H.) is a critical event representing around 2 million strokes worldwide. Hospital confirmations of I.C.H. have substantially increased in the past decade probably because of poor blood pressure levels. Importantly, an early finding of I.C.H. is of critical medical significance in view of the fact that nearly 50% of mortality occurs in the first 24 hours and earlier treatment probably improves results [29].

Computed tomography (C.T.) of the head is the most generally utilised technique for diagnosing intense I.C.H. S.T.A.T. readings are normally interpreted within 60 minutes while routine outpatient studies can take any longer dependent on the accessible radiology staff. Along these lines, the discovery of I.C.H. in routine investigations (particularly those imaged in an outpatient setting) might be fundamentally delayed.

Programmed triage of imaging studies using a machine learning algorithm can recognise I.C.H. earlier, eventually prompting improved clinical results. Such a quality improvement tool could be utilised to naturally deal with the need for translation of C.T. images with assumed I.C.H. and help streamline radiology work process.

4.2.9.10 Robo-Assisted Surgery

Robots empowered with man-made brainpower are increasingly assisting with microsurgical techniques to help reduce professional types that could impact the affected person's recuperation. We understand that a surgeon's capability, especially with new or difficult techniques, typically varies, with ramifications for patient results and fees. Artificial intelligence can both lessen that variability, and help all specialists improve – even the first-class ones. Advanced exams and machine learning of techniques are being applied simultaneously to help demonstrate basic knowledge and exceptional practices from the statistical data associated with the robot-assisted procedure. Machine learning enables surgeons to figure out what is happening during complicated surgical treatment by giving real-time information around the actions the medical professional makes at each stage of the operation. A robot made through Microsure, a business enterprise of the Eindhoven University of Technology and the Maastricht University Medical Center, turned into applied within the operation [29].

The device likewise makes use of A.I. to stabilise any tremors in the surgeon's movements, to ensure the robot correctly performs the surgical operation.

4.2.9.11 Retail Industry

4.2.9.11.1 Recommendation Engines

The sphere of the retailing industry is growing rapidly, therefore the industry needs to analyse data related to their customer to learn his/her issues. Some of the applications of machine learning in the retailing industry are indicated by the behaviour of the customer. The recommendation engine using machine learning techniques predicts the needs of a customer and thereby it recommends a product to a customer. We can observe its application in online marketing platforms like Amazon, eBay, etc. A recommendation engine uses past data like previous shopping experiences, favourite requirements, demographic data, etc. to filter out the best product for its customer. Thus, it helps in the improvement of retail services.

4.2.9.11.2   Market Basket Analysis

Market basket analysis is a modelling technique used by retailers. It is based upon the hypothesis that if you buy a certain group of items you are most likely to buy another group of items. For example, if you buy a pencil you are most likely to buy an eraser or a sharpener. Machine learning uses data mining algorithms to conduct analyses of various previous data. An association is made between products, which increases the selling efficiency of retailers [29].

4.2.9.11.3   Customer Sentiment Analysis

Customer sentiment is the emotional state of a customer during the purchase of a product, whether the customer is happy, neutral or unhappy on purchasing a product. Gaining an understanding of customer satisfaction, loyalty and engagement can prove to be a long-term winning strategy for the industry. The machine learning algorithm can perform a customer-sentiment examination using information received from social networks and online feedback. It is easier to use sentiment analysis on social media. It uses language processing tools to track words to classify positive and negative attitudes of a customer. The algorithm goes through all meaningful layers of speech. The sentiment categories and degrees are recorded and it gives the output as a sentiment rating of customer [29].

4.2.10   FUTURE SCOPE OF MACHINE LEARNING

The scope of artificial intelligence and machine learning are very wide and open. There are a lot of challenges and scope for improvement. Some of them are discussed below.

Advances in artificial intelligence and machine learning can have a significant impact on the drug discovery process. Machine learning can help in finding out the effects of various compositions of drugs on humans. It can be used for reductions in cycle times, costs and labour during research. The use of machine learning techniques in radiology is still evolving. Previous statistics showed that the results of the machine learning system were comparable with those of the radiologists, but the accuracy of the technique can still be improved.

Improvements can be done in the field of personalised healthcare. Using massive data analysis and machine learning, more accurate diagnoses, custom treatments and clinical care plans could be done using some wearable technologies.

Machine learning techniques can also be used to remove language barriers among different people in the world by making systems that can translate every language effectively. Work is already being done in this field but there is much scope for improvement [30].

There is scope for improvement in preventing cybercrime by increasing the level of security over the internet. Work has already been done in this field, but the level of crime still needs to be decreased. This can be done by making a stronger algorithm that can detect fraudulent activities more accurately.

Innovations in machine learning can turn out to be imperative to the agricultural machine vision system. Machine learning technology can be used for detection of weeds, plant diseases and stress detection, yield forecasting, plant water content assurance and soil analysis. Improvements in field operations and monitoring can also be done using machine learning.

Development of self-driving cars, an application of machine learning is still a challenge. Uber and Google are already making a lot of effort in this field, but lots of improvement is needed to be done to ensure the safety of the passengers and public [30].

4.2.11 Conclusions

- Machine learning is an application of artificial intelligence. The system can learn and improve from their experiences without being explicitly programmed.
- Machine learning can be classified as supervised and unsupervised learning. Supervised learning involves a supervisor, whereas in unsupervised learning there is no supervisor.
- There are exceptional varieties of learning for a machine; methods like concept learning, decision tree learning, perceptron learning, Bayesian learning, reinforcement learning and so on.
- An artificial neural network is based on a group of connected units or nodes called artificial neurons, which may loosely be seen as a version of the neurons in a biological mind.
- Deep learning is a specific form of machine learning. It is a learning technique that teaches the machine to make decisions exactly as human beings do.
- The scope of machine learning is very wide, such as data mining, quality management, supply chain management, process planning, the grouping of part families, process modelling, tool condition monitoring, adaptive control, etc.
- Other applications of machine learning include uses in the finance sector, like process automation, security, robot advisors, etc.
- In healthcare industries it is used for cancer diagnosis, identifying tuberculosis, robo-assisted surgery, etc.
- In the retail industry, machine learning is used by recommendation engines, used for market basket analysis, customer sentiment analysis, etc.
- There are a lot of challenges and scope for improvements in the field of machine learning.

REFERENCES

1. T. Ross, The synthesis of intelligence—its implications. *Psychol. Rev.* 45(2) (1938), 185.
2. A.L. Samuel, Some studies in machine learning using the game of checkers. *IBM J. Res.* 3 (1959), 211–229.
3. J. Zhang, F.-Y. Wang, K. Wang, W.-H. Lin, X. Xu, C. Chen, Data-driven intelligent transportation systems: a survey. *IEEE Trans. Intell. Transport. Syst.* 12(4) (2011), 1624–1639.
4. R. Agrawal, T. Imielinski, A. Swami, Mining association rules between sets of items in large databases. ACM SIGMOD Record, No. 22, ACM, New York, NY, 1993, pp. 207–216.
5. M. Meyer, E. Miller, Urban transportation planning: a decision-oriented approach, 2001.
6. L. Tarassenko, *Guide to Neural Computing Applications*, Butterworth-Heinemann, Oxford, UK, 1998.
7. F. Hasson, S. Keeney, H. McKenna, Research guidelines for the Delphi survey technique. *J. Adv. Nurs.* 32(4) (2000), 1008–1015.
8. A.K. Jain, R.P.W. Duin, J. Mao, Statistical pattern recognition: a review. *IEEE Trans. Pattern. Anal. Mach. Intell.* 22(1) (2000), 4–37.
9. R. Polikar, Ensemble-based systems in decision making. *IEEE Circuits Syst. Mag.* 6(3) (2006), 21–45.
10. H.F. Durrant-Whyte, Sensor models and multisensor integration. *Int. J. Robot. Res.* 7(6) (1988), 97–113.
11. F. Castanedo, A review of data fusion techniques. *Sci. World J.* 2013 (2013), 1–19.
12. M.T. Heath, *Scientific Computing*, McGraw-Hill, New York, NY, 2002.
13. F. Marczak, C. Buisson, New filtering method for trajectory measurement errors and its comparison with existing methods. *Transport. Res. Record J. Transport. Res. Board* 2315 (2012), 35–46.
14. P.J. Rousseeuw, A.M. Leroy, *Robust Regression and Outlier Detection*, John Wiley & Sons, New York, NY, 2005.
15. J. Friedman, T. Hastie, R. Tibshirani, *The Elements of Statistical Learning*, Springer, Berlin, 2001.
16. P. Mogha, N. Sharma, S. Sharma. Big Data. *IJRIT Int. J. Res. Infor. Technol.—White Paper* 1 (2013), 223–230.
17. X. Wu, V. Kumar, J.R. Quinlan, J. Ghosh, Q. Yang, H. Motoda, et al. Top 10 algorithms in data mining. *Knowl. Inform. Syst.* 14(1) (2008), 1–37.
18. R.O. Duda, P.E. Hart, D.G. Stork, *Pattern Classification*, John Wiley & Sons, New York, NY, 2012.
19. E. Alpaydin, *Introduction to Machine Learning*, MIT Press, Cambridge, MA, 2014.
20. L. Breiman, J. Friedman, C.J. Stone, R.A. Olshen, *Classification and Regression Trees*, CRC Press, Boca Raton, FL, 1984.
21. W.S. McCulloch, W. Pitts, A logical calculus of the ideas immanent in nervous activity. *Bull. Math. Biophys.* 5(4) (1943), 115–133.
22. M. Dougherty, A review of neural networks applied to transport. *Transport. Res. Part C Emerg. Technol.* 3 (4) (1995), 247–260.
23. M.G. Kraft's, E.I. Vlahogianni, Statistical methods versus neural networks in transportation research: differences, similarities and some insights. *Transport. Res. Part C Emerg. Technol.* 19 (3) (2011), 387–399.
24. P.J. Werbos, Backpropagation through time: what it does and how to do it. *Proc. IEEE* 78(10) (1990), 1550–1560.
25. V. Vapnik, *The Nature of Statistical Learning Theory*, Springer Science & Business Media, New York, NY, 2013.
26. R. Fletcher, *Practical Methods of Optimization*, John Wiley & Sons, New York, NY, 2013.
27. K. Zhu, H. Wang, H. Bai, J. Li, Z. Qiu, H. Cui, et al., Parallelizing support vector machines on distributed computers. *Adv. Neural. Inf. Process. Syst.* 20 (2008) 257–264.
28. C.-C. Chang, C.-J. Lin, LIBSVM: a library for support vector machines. *ACM Trans. Intell. Syst. Technol.* 2(3) (2011), 1–27.

29. T. Razzaghi, I. Safro, Scalable multilevel support vector machines. *Proc. Comp. Sci.* 51 (2015) 2683–2687.
30. C.C. Aggarwal, C.K. Reddy, *Data Clustering: Algorithms and Applications*, CRC Press, Boca Raton, FL, 2013.

4.3 SOFTWARE DEVELOPMENT FOR INDUSTRY 4.0

Lokesh Singh, Sushil Kumar Maurya, Ashish Das, and K. Jayakrishna

4.3.1 INTRODUCTION

Various types of software are used in manufacturing industries, such as product development process, product data management, product life-cycle management, E.R.P., computer-aided design, computer-aided manufacturing, computer-aided engineering, etc. In the competitive manufacturing industry, the software focuses on benefits in the manufacturing industry which are as follows:

- Automated manufacturing
- Reduces costs
- Enhances internal and external communication
- Improves customer satisfaction
- Designing and material analysis.
  (1) **Automated Manufacturing**
      Software has a major impact in the automated manufacturing sector that increases mass production, saves time and reduces defects effectively. Manpower is reduced on a large scale when producers shift their focus from manual manufacturing to automated manufacturing that makes it more precise and simpler, so automated manufacturing increases productivity.
  (2) **Reduces Costs**
      Software together with the operations inside the production area substantially reduces costs. Enterprise resource planning systems allow monitoring of work operations which avoid re-entry, ensure excellent accuracy and reduce the margin of error on the production. So, software reduces costs and improves efficiency.
  (3) **Enhances Internal and External Communication**
      Software inside the production area allows producers to share data associated with production easily, not only with internal employees, but also with outside stakeholders including dealers, suppliers and end-consumers. Hence, collaboration and communication with business partners improve immensely. So, software programs for the manufacturing industry allow the central contributors of numerous departments to coordinate properly and improve overall decision-making.
  (4) **Improves Customer Satisfaction**
      In production, it's important for products to be supplied to the customers on time and reduce late supplies. Many important factors assist manufacturers to ensure on-time supply, including optimum inventory

control, precise manufacturing planning, control of supply channels and retaining control as the scheduling processes. Software provides on-time delivery of products and improves customer satisfaction.

(5) **Designing and Material Analysis**

Software for the manufacturing industry fulfils the design functions through products that are modelled in software together with all their dimensions. Software minimises the manual determination of the measurements and ensures high levels of precision. Materials analysis is used for the selected design for a particular material through analytical software. So, software provides a better design in less time and no waste raw materials.

In the manufacturing industry, product creation is rarely process-driven. In this industry everyone has a specific set of responsibilities that they follow, such as an employee tightening a screw all day. Every task follows to the next challenge in the process that is finished by the specialist. Parts pass through all processes and the output is the final product. It works well and companies achieve 99.996% defect-free products. In the manufacturing process a finished product can be precisely created. Suppose your company is contracted to create software that will search millions of images to find the faces of criminals. Despite the heavy process-driven environment, the developer must engage in creating things. A good manufacturing process can be executed without thought by a worker. Software development is not a process, but the creation of processes to be executed by a computer. Software engineering has totally understood the nature of processes. A process is a set of steps. If something can be expressed as a set of steps, it can be automated. Automation cannot perform original creation; therefore all processes are orthogonal creation. Fujitsu software technologies have improved productivity using the Toyota production system. Agile software development and the store management method were introduced as the basic concepts of T.P.S. (elimination of muda (waste), heijunka (production) and jidoka (automatic detection of abnormal conditions)) in the I.T. software field. Visual management through agile development, store management and the lean manufacturing method are practical techniques which are useful as the fundamental manufacturing software that increases production rates and improve quality in a short period. Industry 4.0 is characterised as the co-ordination within automated systems, connected and networked with global systems of robotics and automation, and internet of things is the new and emerging foundation that is giving rise to an upsurge of human ingenuity and creativity. But at the same time, mechanical engineering is also facilitating a quantum leap to higher levels of opportunities to address the noble objectives of zero-defect engineering and zero-waste manufacturing, and at the same time making mankind capable of achieving the Millennium Goals of sustainability and green manufacturing.

C.A.D. systems have been applied in various facets of engineering. The main scope is to increase the productivity of the designer, as well as improving the quality of the design. A lot of C.A.D. programs make possible the creation of three-dimensional models, which can be viewed from different directions. Computer-aided design software available on the market is widely used due to improving the productivity of the designer by facilitating various features that reduce product development time. However, it has been observed that most of the manufacturing industries frequently design and manufacture similar types of components and assembly often uses standard parts. If a designer is provided with custom programs for parts and assembly, he only has to input values for key parameters and the part will automatically be generated. Almost all C.A.D. software allows customising to make it suitable for your needs. Software will have a standard library and it can be customised for generating solid standard parts like washers, equal angles and ball-bearings. This customised software automates the iterative process of modelling of various types of washers and structural sections. The time spent by the designer in routine work is thus drastically saved, and he can use this time in creative aspects of the design process.

In today's dynamic commercial enterprise structures, Mr Narendra Modi as the Prime Minister of India had released the 'Make in India' application to post India on the world map as a manufacturing hub and provide global popularity to the Indian economy. India is predicted to be growing the fifth largest manufacturing country in the world by the end of 2020. 'Make in India' leads the way to becoming the hub of global hi-tech production together with GE, Siemens, HTC, Toshiba and Boeing who have set up production plants in India which access India's market of more than billion customers and growing purchasing power.

Today, lean manufacturing is used in global software enterprise industries through which operational costs are reduced and productivity increased. Increasing productivity aims to develop better software that continuously delivers customer requirements. The lean method was found to be successful in the automobile industry. During the last decade, the lean technique has had a major impact on the competitiveness of companies through enhancements of their process efficiency and reduction of their waste. The main advantages of the lean approach are awareness of the flow of work, purchaser satisfaction, improved quality and supply lead time, earlier responses and decreases in customer mistake reporting and improved communication between stakeholders. Manufacturing has emerged as one of the high growth sectors in India.

The fourth industrial revolution is focused on mechanisation and data exchange in manufacturing technology for the future of urban and rural industries in India. Industry 4.0 creates modular structures in smart industry that consists of cyber-physical systems, artificial intelligence, I.o.T. and cloud computing. The digital M.S.M.E. scheme of the Ministry of Micro, Small and Medium Enterprises moves towards new approach i.e., cloud computing for digital M.S.M.E. approved by their production and business method with improved quality of products and services. Digital M.S.M.E. is a 'pay as you use' model so the manufacturer does not have to invest upfront. Software is not the manufacturer's responsibility and there is flexibility in choice of services. It is more affordable than the market rate and has higher security of data and accessibility from anywhere at any time (Figure 4.20).

4.3.2 History of Software in Manufacturing Industries

The first computer in the modern age is generally found to be E.N.I.A.C. which was developed by the U.S.A. at the end of the Second World War; however, the concept of

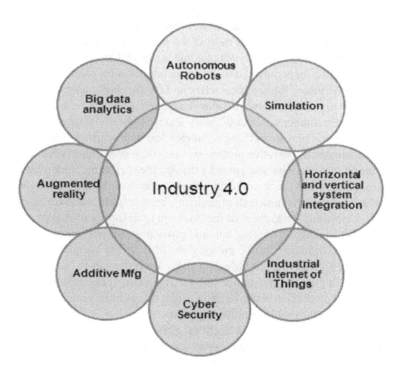

FIGURE 4.20   Technologies for Industry 4.0 [9].

software developed in the 19th century. The development of operating systems considered to be software development occurred in the late 1960s. In December 1968, IBM 'unbundled' its software from its hardware, which constituted a revolution in software and software development. By the mid-1970s, many software program suppliers established the first software program packages.

Software was changing from a cottage industry to the mass manufacturing model. Many industries came to supply applications such as economic and production packages. Software application led to the development of programming languages and the standardisation of operating systems, but data management was a crucial issue. During the 1950s and the early 1960s, there had been no standardised manner of accessing data and each program needed to manage its own. The records might be stored in tables of related rows and columns which were called R.D.B.M.S. (database management system). This opinion led to a revolutionised I.T. industry. The second most important method became the growth in E.R.P. systems. E.R.P. basically means the synthesis of high applications, generally utilised in manufacturing or accounting systems.

Effective C.R.M. and E.R.P. software for manufacturing industry can play an integral part in this process for a manufacturing business. Much software is developed for the manufacturing industry, including just-in-time (J.I.T.), lean manufacturing, rapid manufacturing, flexible manufacturing and agile manufacturing. Enterprise Resource Planning (E.R.P.) is a business method management software program that permits any organisation to apply a system of integrated application. E.R.P. software

integrates all aspects of an operation including: product planning, evolution, production, sales and marketing. Resource planning software offers a manufacturing
system for make to order and make to stock as well as customisations (Table 4.3).

4.3.3 NEED FOR SOFTWARE DEVELOPMENT IN INDUSTRIES?

Software development is the fastest-growing technology in the I.T. industry and is
associated with software application for devices such as smartphones, tablet, computers, digital enterprise and portable internet. The software industry includes businesses for development, maintenance and publication of software that are using
different business models, mainly either 'maintenance-based or cloud-based'. The
industry also includes software services such as training, documentation and data
recovery. Industrial software development needs good domain understanding in
order to build effective software whether it is a simple diagnostic and service tool for
a set of actuators or designing the enterprise as a cloud-based solution to access and

TABLE 4.3
History of Software in Manufacturing Industries [6]

Sr. No.	Software	Founded Year	Features
1	Computer-aided manufacturing (C.A.M.)	1950	Computer-controlled machinery to operate a manufacturing process.
2	Computer-aided engineering (C.A.E.)	1970	To simulate performance for enhanced product designs.
3	S.A.P.	1972	System analyses and program run, started by former IBM employees who were the first company to create enterprise software to manage business operation and customer service.
4	Computer-aided design (C.A.D.)	1974	To facilitate the generation, modification and optimisation of a part or a compilation of parts.
5	Oracle	1977	Larry Ellison developed Oracle, a database program.
6	Product development process (P.D.P.)	1980	Identify a customer need – idea generation for product development.
7	Product data management (P.D.M.)	1985	Capturing and keeping data on products and services through their development.
8	Enterprise resource planning (E.R.P.)	1990	E.R.P. uses a single database containing all the data that keeps the processes running smoothly, ensuring visibility, accessibility and consistency.
9	Product life-cycle management (P.L.M.)	1998	Managing the entire life-cycle of a product from inception through engineering design and manufacture to service and disposal of manufactured products.
10	Customer relationship management (CRM)	2004	Improve customer retention with improve profitability.

visualise critical information anytime and anywhere from any type of device. The software development is changing drastically, making more and more new technologies and tools readily available and allowing for faster release cycles. Those faster cycles have come about in response to online users. The future of work for software developers lies in the industry's ability to embed diversity, intelligence and creativity into product teams, in order to prioritise a user perspective. Software tends to get commoditised and one requires expertise to operate at higher levels of abstraction. Some of the software technologies that are relevant today include

- Big Data infrastructure and data mining.
- Distributed high-availability systems
- Development for mobile computing
- Development for the web
- Data science
- High-performance applications/high-performance computing
- Search technologies
- Application and network security
- Enterprise systems
- Networking systems [4, 5].

4.3.4   VISION FOR SOFTWARE DEVELOPMENT FOR INDUSTRIES

Today, the competitive global market demands high-quality products with accuracy and precision. For the highly competitive environment in production, the software system is most important for design, testing, process planning, manufacturing and assembly. Software was developed for data visualisation, analysis and watch circle for monitoring automated quality control. Software as a tool for design assignments is responsible for reducing the product development cycle time. Software provides customisation facilities which are made more specific according to industry requirements. To become a prime performer in the global marketplace a company must provide highly innovative software designs, software development and internet marketing services that will drive users towards growth. Worldwide reputation is the dream of every company and they want to achieve it through their work. The following points for software development of industries are mentioned below:

- To improve the software by exploring innovative solutions for customers
- To become an integrated I.T. solutions provider
- To complement India's literacy drive and hence contribute globally
- To build the highest standards of reliability and cost containment
- To offer seamless service and build value for users
- To be a process-driven, professionally managed and highly profitable organisation.

4.3.5   COMPARISON OF PAST AND PRESENT SCENARIO OF SOFTWARE IN INDUSTRIES

India is the world's largest democracy and the second-fastest growing economy. India has grown to become a trillion-dollar economy characterised by foreign investment and direct policy. The software industry offers software product/packages. Software is a knowledge-driven industry. It requires a team of highly skilled professionals for its success. Today, the software is continuously expanding in India. It is a leading destination for I.T. Software is the leading industry in India as the gross revenue has grown from 1.2% from 1997–1998 to 5.8% from 2015–2018. The software solutions industry is a major revenue earner. The B.P.O. (Business Process Outsourcing) and K.P.O. (Knowledge Process Outsourcing) industry is surging, with more companies looking to offshore their customer service departments. There are newer areas that are emerging that are making India a potential winner for investors. Cloud computing is broken down into three segments: application, storage and connectivity. Each segment serves a different purpose and offers different products for businesses and individuals around the world. In June 2018, a study conducted by VersionOne found that 91% of senior I.T. professionals don't actually know what cloud computing is, and two-thirds of senior finance professionals are not clear on the concept, highlighting the young nature of the technology. Cloud computing is a technology that uses the internet and central remote servers to maintain data and applications. Cloud computing allows consumers and businesses to use applications without installation and to access their personal files at any computer with internet access. This technology allows for more efficient computing by centralising storage, memory, processing and bandwidth. Cloud computing as a technology has existed for quite a long time now and the way enterprises have used it has been rapidly changing. It began as a simple hosting service and has evolved to I.a.a.S., P.a.a.S. and S.a.a.S. Looking at the present-day scenario, large-scale cloud computing is now accessible to everyone. It's not just limited to big players in the market and now is open to anybody, anywhere. Cloud has now grown into a well-known approach for the management and deployment of applications for a growing number of businesses.

Today, technology drives a revolution in the way software development has been working for a long time. New terms have taken its place such as I.o.T., Big Data, data science and artificial intelligence which have become a simple reality. Operations software is not the need, the need is much more. Industries are looking for ways to reduce costs, and technology to support them in all possible ways. The I.T. industry has to gear up quickly in order to respond to the needs of the current generation customer. The present has to predict the future by connecting the dots based on the events happening within the industry which can impact your business. The industry can see the opportunity to innovate and survive [7, 8].

During the late 1990s and early 2000, India was seen as a place to invest in the field of I.T. (Information Technology), due to its population and the added advantage of language over its neighbouring nation, China. There was a large demand for people with any engineering background to their computer skills. The I.T. industry hired them by thousands and trained them for years to make them productive. During this period, technology was seen as an enabler to automate the operations of an organisation, allowing them to reduce manual work and think productively to improve their business. Thus, the I.T. industry developed systems to manage the operation of an organisation, to help them save time and reduce costs in their day-to-day activities. During this phase, there were a lot of areas which could not be automated, as there were no solutions to many of the problem statements of different industries. Less

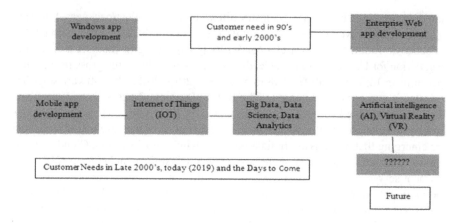

FIGURE 4.21 Software development during the fourth industrial revolution [11].

than 20% of the software went live and was used by the customer and the remainder could not be delivered because of quality issues, changes in customer requirements at the time of delivery, or inability to deliver as per the deadlines. This has led to various improvements and research in the field of I.T. (Figure 4.21).

4.3.6 EXPECTING FUTURE SOFTWARE DEVELOPMENT IN INDUSTRIES 2050

Software development is the work of solving large and complex programs in a cost-effective manner. The development and maintenance of software products have become important criteria. In the early years, engineers faced many problems without having better knowledge in this field, such as late delivery of software, poor quality, user requirements that were not completely supported by the software, difficult maintenance, unreliable software and the lack of a systematic approach. Nowadays, this problem has been rectified and the advanced development process and standards are implemented. But software developers and the software industries still face many problems. The focuses of future scenarios for the software developed in India is improving the quality of the software. In 2050, machines will write the software and test the software. There will be 10% of the requirements for humans to write the software, as nowadays machine learning and artificial intelligence is trending to automate every device or application. So, in 2050 there will be less need for humans in the software industry. Digitisation and technologies like the internet of things, artificial intelligence, 3D printing, machine learning and robotics not only disrupt the ways we design, produce, manage and maintain products and services at a fast pace, but also the way we will work in the future. Machine learning and artificial intelligence will play an increasingly central role in the field of software development. Blockchain technology holds incredible potential for many industries, especially when used in tandem with internet of things (I.o.T.) data, artificial intelligence (A.I.) and fog computing. Software developers will be focused on new solutions that leverage Blockchain ledgers such as solutions to enable micropayments and smart contracts or end counterfeiting in the supply chain. After a few years, developers will be working mostly on integration between different services instead of developing a custom software solution. The next trend will be programming human behaviour – creating computational models of human behaviour and developing algorithms to aid the customers/users with possibilities and choices. Finding the trends using digital behaviour can calculate the next move of the user. Programming perceptual processes will be the next big thing in software development, and it will help mediate digital identity and behaviour. The evolution of DevOps will include security testing earlier and at more points in the development. Security testing is currently a bottleneck for delivery, and the cost is highest to correct code when done late in the cycle. The developers will have real-time feedback on the security of the code. A.I. and machine learning-driven product features are already an integrated part of software development for e-commerce, movie watching and social media. Now A.I.-first software, from conversational virtual assistants to self-driving technologies, are becoming mainstream in software development. Software platforms that generate objective ratings for products and services, based on an analysis of actual prior usage data, are going to be critical in enabling better decision-making.

4.3.7 METHOD USED FOR SELECTION SOFTWARE IN INDUSTRY

There are many software development methodologies, so you can choose those that work best for the project. All methodologies have different strengths and weaknesses. An overview of the most commonly used software development methodologies follows below.

4.3.7.1 Waterfall Development Methodology

The waterfall method is considered the traditional software development method. It's a rigid linear model that consists of sequential phases (requirements, design, implementation, verification, maintenance). The distinct goals that are accomplished in each phase must be 100% complete before the next phase can start, and traditionally there is no process for going back to modify the direction. This method makes it easy to understand and manage with stable requirements. It is often slow and costly due to the rigid structure (Figure 4.22).

4.3.7.2 Rapid Application Development Methodology

Rapid application development (R.A.D.) is a condensed development process that produces a high-quality system with low investment costs. This R.A.D. process allows our developers to quickly adjust to shifting requirements in a fast-paced and constantly changing market. The rapid application development method contains four phases: requirements, planning, user design and construction. The user design and construction phases are repeated until the user confirms that the product meets all requirements. R.A.D. is the most effective method with a well-defined business objective and a clearly defined user group. It is especially useful if the project is of small to medium size and is time-sensitive. It requires a stable team composition with highly skilled developers and users who are deeply knowledgeable about the application area (Figure 4.23).

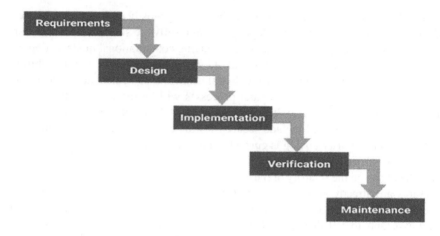

FIGURE 4.22   Waterfall development methodology [1].

FIGURE 4.23   Rapid application development methodology [2].

4.3.7.3   Agile Development Methodology

Agile methods attempt to minimise risk (such as bugs, cost overruns and changing requirements) when adding new functionality by developing the software in iterations that are mini-increments of the new functionality. The benefit of multiple iterations is that it improves efficiency by finding and fixing defects and expectation mismatches early on. Agile methods use real-time communication which fails to provide new users with documentation to get up to speed. They require a huge time commitment from users and are labour-intensive because developers must fully complete each feature within each iteration for user approval. Agile methods are similar to R.A.D. and can be inefficient in large organisations. Programmers, management and organisations accustomed to the waterfall method may have difficulty adjusting to agile, so a hybrid approach often works well for them (Figure 4.24).

4.3.7.4   DevOps Deployment Methodology

DevOps deployment is centred on organisational change that enhances the collaboration between the departments responsible for different segments of the development life-cycle such as development, quality assurance and operations. DevOps is focused on improving time to market, lowering the failure rate of new releases and the lead time between fixes and prioritising minimal disruption and maximum reliability. DevOps aims to automate continuous deployment to ensure everything

FIGURE 4.24 Agile development methodology [2].

FIGURE 4.25 DevOps deployment methodology [2].

happens smoothly and reliably. The benefit of DevOps is reducing the timeframe and improving customer satisfaction, product quality and employee productivity and efficiency (Figure 4.25).

4.3.8 AVAILABLE SOFTWARE FOR DIFFERENT AREAS IN INDUSTRIES

Software is an incredibly in-demand and rewarding field to be a part of in today's competitive job market. You have more opportunities to work on various types of software for different areas in industries.

4.3.8.1 Industrial Design Software

4.3.8.1.1 Adobe Photoshop and 3D

Photoshop is an industry standard in digital imaging and is used by professionals around the globe in fields from video-editing to design. New features and upgrades to Photoshop are making it a tool of choice for industrial designers who are looking for a tool with impeccable 3D capabilities. Photoshop CC allows designers to add dimensions to their designs using its rich set of 3D image creation, editing and printing tools.

4.3.8.1.1.1 Key Features
- Visualise 3D designs in the real world
- Easily print to a 3D printer or 3D print service from directly within Photoshop CC
- Automatic mesh repair and automatic generation of support structures and rafts

- Includes all the 3D editing tools that were previously available in Photoshop Extended.

4.3.8.1.2   AutoCAD

Autodesk's AutoCAD is an industrial design tool for Windows or Mac. It promises to help industrial designers 'document with confidence', because it enables them to create designs and speed documentation work with its built-in productivity tools. With AutoCAD, industrial designers have the ability to work across integrated desktop, cloud and mobile solutions.

4.3.8.1.2.1   Key Features
- Clearly see details with Line Fading and smoother curves
- Automatically create appropriate measurements based on your drawing context with Smart Dimensioning
- Tools are designed for maximising productivity
- Design and visualise nearly any concept with 3D free-form tools.

4.3.8.1.3   SolidWorks

Dassault Systèmes' SolidWorks is a 3D C.A.D. design software tool that is a perfect solution for industrial designers. With SolidWorks, industrial designers can fast-track their designs through manufacturing, thanks to the new user interface, new tools and new process enhancements that enable users to complete work faster and more easily than ever before.

4.3.8.1.3.1   Key Features
- Fewer 'picks and clicks'
- Increased modelling flexibility
- More intuitive interface
- Easier access to commands
- Communicate, collaborate and work concurrently across teams, disciplines, customers and vendors with mechatronic design, concurrent design and streamlined electrical/mechanical design
- Innovative design simulation for more efficient analysis to solve complex problems and discover potential errors before they occur.

4.3.8.1.4   Fusion 360

Autodesk's Fusion 360 is 3D C.A.D./C.A.M. software for product design. In fact, Fusion 360 is the first 3D C.A.D., C.A.M. and C.A.E. tool of its kind. It connects your entire product development process in a single cloud-based platform that works on both Mac and P.C. With this single industrial design tool, users can design, test and fabricate, work anywhere and collaborate with anyone.

4.3.8.1.4.1   Key Features
- Quickly iterate on design ideas with sculpting tools to explore form and modelling tools to create finishing features
- Test fit and motion, perform simulations, create assemblies, and more

- Create tool paths to machine components or use the 3D printing workflow to create a prototype
- Join together design teams in a hybrid environment that utilises the power of the cloud as needed and uses local resources when needed.

4.3.8.1.5   Siemens NX for Design

NX for Design is Siemens' P.L.M. software. This industrial design software tool is an integrated product design solution that 'streamlines and accelerates the product development process for engineers who need to deliver innovative products in a collaborative environment'. NX for Design is a step ahead of C.A.D.-only solutions because it offers an advanced level of integration between development disciplines to create an open, collaborative environment. The NX for Design solution is specifically for industrial design and styling and is a powerful tool that enables users to optimise form, fit, function and user experience.

4.3.8.1.5.1   Key Features
- Freeform shape modelling
- Reverse engineering
- Flexible, robust computer-aided industrial design and styling software for accelerating product engineering through fast concept design and modelling.

4.3.8.2   Information Technology Industry

A wide variety of software comes under the domain of the information technology industry. Some of this software is as follows:

4.3.8.2.1   Database Design and Development

Database design is a prerequisite for working efficiently with databases, as it reduces maintenance effort, minimises the chances of errors in tasks, provides a mechanism for a good communication channel among the users and overall reduces the time taken and efforts made by the users to perform a task.

4.3.8.2.1.1   Phases of Database Designing   Database designing includes a lot of phases to get it completed in an effective way. Database designing is performed in connection with the application design.

- Collecting requirements and their analysis
- Conceptual design
- Logical design
- Moving towards normalisation process
- Physical design

Some of the most well-known database software programs include:

- Microsoft Access
- Microsoft Excel
- Microsoft SQL Server

- MySQL
- Oracle RDBMS
- SAP
- Teradata.

4.3.8.2.2 Networking

Networking is a general phrase for software that is designed to help set up, manage and/or monitor computer networks. Networking software applications are available to manage and monitor networks of all sizes, from the smallest home networks to the largest business networks. Networking offers numerous benefits including on-demand provisioning, automated load balancing, streamlined physical infrastructure and the ability to scale network resources with applications.

4.3.8.2.2.1 SolarWinds Network Performance Monitor
SolarWinds Network Performance Monitor is easy to set up and can be ready in no time. The tool automatically discovers network devices and deploys within an hour. Its simple approach to oversee an entire network makes it one of the easiest to use and it has a most intuitive user interface. The product is highly customisable and the interface is easy to manage and change very quickly.

4.3.8.2.2.1.1 Key Features
- Automatic network discovery and scanning for wired and Wi-Fi computers and devices
- Quickly pinpoint issues with network performance with net path
- Easy to use performance dashboard to analyse critical data points and paths across your network.

4.3.8.2.2.2 PRTG Network Monitor
PRTG Network Monitor software is commonly known for its advanced infrastructure management capabilities. All devices, systems, traffic and applications in your network can be easily displayed in a hierarchical view that summarises performance and alerts. PRTG monitors I.T. infrastructure using technology such as S.N.M.P., W.M.I., S.S.H., flows/packet sniffing, H.T.T.P. requests, R.E.S.T. A.P.I.s, Pings, S.Q.L. and a lot more. It is one of the best choices for organisations with low experience in network monitoring.

4.3.8.2.2.2.1 Key Features
- Its ability to monitor devices in the data centre. The user interface is very easy to use
- It is used to scan the code and a summary of the device is displayed on the screen.

4.3.8.2.3 Material Handling and Purchasing Software

4.3.8.2.3.1 E.R.P. Software
Enterprise Resource Planning (E.R.P.) is business management software that allows an organisation to use a system of integrated applications to manage the business. E.R.P. systems are large computer systems that integrate application programs in accounting (i.e., accounts receivable), sales (i.e.,

FIGURE 4.26 Process of E.R.P. System [10].

order booking), and manufacturing (i.e., product shipping) and the other functions in the firm, this integration is accomplished. E.R.P. systems are the use of databases to store data for various system modules. Material base data is typically imported from the E.R.P. system and used for inventory handling in manufacturing. The key functional components handle identification of material, tracking of consumed material, tracking of produced material, control of source and destinations for material and material compatibility weighing. The ECS Material Management Module also provides powerful product genealogy reporting (Figure 4.26 and Figure 4.27).

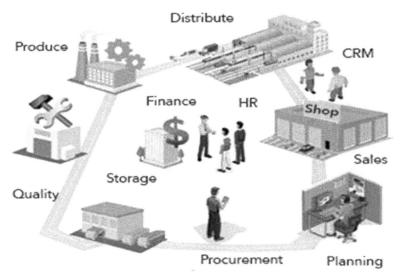

FIGURE 4.27 Manufacturing industry for E.R.P. [10].

4.3.8.2.3.1.1   Key Features
- Tracking and tracing of individual parts, lots, batches, assemblies
- Material identification, real-time reporting of material consumed and material produced
- Management of material compatibility and availability
- Weigh and dispense support
- Material handling.

4.3.8.2.3.2   CRM Software   Customer relationship management (C.R.M.) is a model for managing a company's interaction with current and future customers. It involves using technology to organise, automate and synchronise sales, marketing, customer service and technical support. C.R.M. is a business strategy that aims to understand, predict and manage the needs of an organisation's current and potential customers. C.R.M. is concerned with the creation, development and enhancement of individualised customer relationships with carefully targeted customers and customer groups in maximising their total customer life-time value.

4.3.8.2.3.2.1   Key Features
- It helps a business to keep customers
- It helps the business to understand what it needs to do to get more customers
- It reduces costs by managing costly complaints and finding out what services are useless for customers
- It helps a company figure out if its product is working and increases profit
- It manages customer relationships (Figure 4.28).

Advantage

- It has easy integration with E.R.P./M.E.S., etc.
- It is easy to add new protocols and processes
- It is possible to communicate between components on the shop floor
- It should boost production flexibility, enabling a facility to rapidly adapt its operations to market changes
- It facilitates the interaction between humans and technology by providing information at all industrial plant levels.

Limitation

- Reliability and stability are required for machine-to-machine communication
- It requires maintaining integrity of production processes
- It requires educating staff to adopt this fourth industrial revolution.

4.3.9   Summary

This chapter focuses on software development for Industry 4.0 (smart factories) that takes a look at the future of process and data management. Software is an effective and efficient business transformation in the era of I.o.T., digitalisation and cloud

FIGURE 4.28 Customer relationship management [3].

computing. The outcome of this chapter is a software architecture that serves a real production plant optimisation which integrates several factories, sites and plants in order to have symbiotic effects on a large scale. This leads not only to smart factories, but also to a smart industrial park. Industry 4.0-aware business process management is an important part which transforms them into a software design. This concept includes a decentralised monitoring system using I.o.T. (internet of things), a machine learning component and a decision support system with A.I. (artificial intelligence), cloud-based and mobile client applications for the end users and system architecture so that new components can be integrated easily. Machine learning and A.I. (artificial intelligence) lead to recommendations for cost reduction and quality improvement. So, the system will measure, visualise and simulate all processes and data flows, and forecast the production cost and output on a very fine granular level [12].

4.3.10 CONCLUSION

This chapter discusses the software development of a fourth industrial revolution. As per requirements that the industry adapt to the industrial internet of things, artificial intelligence, cloud computing and machine learning, there is an increasingly important role for software quality. Further developing the software in the fourth-generation industrial revolution faces the problems of finding ways to solve the problem in the current and future software scenario in industries.

REFERENCES

1. CMMI Standard in software development process. www.ijcns.com/pdf/ijcnsvol5no1-2.pdf.
2. Software development methodologies. https://www.synopsys.com/blogs/software-security/author/synedt/.
3. CRM: Manufacturing smarter customer relationships. https://www.themanufacturer.com/articles/manufacturing-smarter-customer-relationships/.
4. Eduardo, T. Katayama, and Alfredo Goldman. 2011. "From manufacturing to software development a comparative." *International Conference on Agile Software Development LNBIP*, 77, 88–101.
5. Padberg, Frank. 2000. "Estimating the impact of the programming language on the development time of a software project." *Proceeding International Software Development and Management Conference.* ID:32432000 S.287–298.
6. Alberts, G., A. van den Bogaard, and F. Veraart. 2005. "History of the software industry: The challenge." CWI ISSN 1386-369X.
7. Dwivedi, A. 2013. "Role of computer and automation in design and manufacturing for mechanical and textile industries: CAD/CAM." *International Journal of Innovative Technology and Exploring Engineering*, 3, 8.
8. Paliotta, John. 2015. "Software quality and industry 4.0: The industrial internet of things." https://dzone.com/articles/software-quality-and-industry-40-the-industrial-in.
9. Cheng, Guo-jian, and Li-Ting Liu. 2016. "Industry 4.0 development and application of intelligent manufacturing." *International Conference on Information System and Artificial Intelligence (ISAI).* doi:10.1109/ISAI.2016.0092.
10. Nandi, M., and A. Kumar. 2016. "Centralization and the success of ERP implementation." *Journal of Enterprise Information Management*, 29(5), 728–750. doi:10.1108/JEIM-07-2015-0058.
11. Senthilmurugan, C. 2016. "The scenario of software industries in India." *International Journal of Technology and Engineering System (IJTES)*, 8, 158–163.
12. Richard, E. Crandall. 2017. "Industry 1.0 to 4.0: The evolution of smart factories." http://www.apics.org/apics-for-individuals/apics-magazine-home/magazine-detail-page/2017/09/20/industry-1.0-to-4.0-the-evolution-of-smart-factories.

5 Sustainable SMART Factories
Compliance with Environmental Aspects

5.1 MONITORING MANUFACTURING PROCESS PARAMETERS FOR NEGATIVE ENVIRONMENTAL IMPACTS: CASE STUDY FROM COLOMBIA

J. Martinez-Girlado and K. Mathiyazhagan

5.1.1 INTRODUCTION

The Industry 4.0 system can generate a regular flow of information much better than the use of logistics strategies and traditional production systems. In addition, the information is exchanged very quickly, either internally or externally. This allows for possibilities with external logistics actors, since it could easily allow adaptations to changing situations, both internally in the industrial plant or production chain, and at a general level (World Economic Forum, 2017).

Industry 4.0 also aims to respond to current problems related to economics by managing natural resources and energy. With a system based on a communications network and instant and permanent exchange of information, there is greater preparation to achieve correspondence with the needs and availability of each element of the system, allowing improvements and gains for productivity as well as in the economy of the resources. However, due to the short time since the concept was proposed in Germany and its proliferation, it is still necessary to investigate the long-term environmental and social impacts.

According to Mashhadi and Behdad (2017), intelligent manufacturing in an Industry 4.0 environment requires the development of infrastructures for detection, wireless communications, production systems and information traceability. Because of this, in intelligent infrastructures, life-cycle assessment (L.C.A.) techniques are unable to determine the true impact. Therefore, there is a gap between the Industry 4.0 technologies and the system to evaluate their environmental impact.

Colombia is an excellent example of how the technological revolution needs to be guided. Colombia is a developing country with 48 million habitants. Even though the main economic source is agricultural vocation and some agri-industries such as plantain, coffee and citrus, Colombia plans to make an incursion into some Industry 4.0 tools. This is due to explosive demographic and industrial development, especially in the communications and manufacturing sectors. Furthermore, the government is trying to unify its enterprises with a new manufacturing Industry 4.0 system. (DANE, 2018).

In consequence, a conflict between economic growth and environmental concerns has been raised in Colombian cities. Supporting economic growth through enterprises policies and benefits generates a negative impact on the environment. To manage this situation, several methodologies have been suggested to measure environmental development in order to find practices generating more economic benefit and less pollution. Nevertheless, the government has not proposed any yet. Thus, environmental Colombian institutions are trying to measure the environmental impact of Industry 4.0 with laws and regulations related to traditional processes.

Thus, there emerge methodologies to maximise productivity and reduce consumption of natural resources. Environmental Performance Indicators (E.P.I.s) are a way to measure, compare and analyse how changes in manufacturing, especially changes related to an Industry 4.0 impact the environment. Furthermore, the gap between the Colombian Industry 4.0 and the environmental regulations is exposed.

5.1.2 Environmental Impact Measurement

The performance measurement looking for environmental sustainability has been a major problem for companies. In many cases, these are the minimum standards. According to Hart and Milstein (2003), companies have to become proactive and extend the normative and independent control mechanisms, trying to promote effective sustainability that encompasses the well-being and needs of different stakeholders.

The pro-active activities proposed and implemented need to be measured adequately. This means using an appropriate set of indicators (Giljum et al., 2011). This measurement has several advantages, among them, the characterization of costs associated with environmental practices and the evaluation of their efficiency (Diabat & Govindan, 2011; Zhu et al., 2008), as well as economic, environmental and social responsibility benefits, differentiation in the market and others granted in the regulations (Eltayeb et al., 2011). For this reason, it is important to find a strategy to deal effectively with sustainable performance on companies (Atkinson et al., 1997; Neely et al., 2002; and Epstein, 2008).

There are several definitions of Environmental Performance Indicators or E.P.I.s. Henri and Journeault (2008) state that E.P.I.s represent numerical measures. These measures provide crucial material on environmental impact and compliance with internal and external policies. A more complete definition is provided by Hourneaux et al. (2014), who define E.P.I. as an interior process and a management tool to determine if the environmental performance is according to established ranges. E.P.I.s represent qualitative and quantitative measurements.

Currently, exist several E.P.I.s in the literature consulted, which involve different measurement approaches depending on who is doing the environmental measurement, i.e., state agencies or academics (Petrovic et al., 2012). Some indicators of state and non-governmental agencies have been developed by recognised organisations such as the United Nations, the World Bank, the World Resources Institute (W.R.I.) and the World Business Council for Sustainable Development (O.E.C.D.).

5.1.2.1 Functions and Characteristics of Composite Indicators of Environmental Performance

E.P.I.s have been considered for the evaluation and control of the environmental performance of companies. Due to the reliability of its information, managers have a strong information base to carry out the decision-making process, allowing the identification and improvement of the critical points of the organisations (Medel-Gonzales & García, 2012).

One function of environmental performance indicators is measuring the environmental performance achieved. Also, they establish corrective actions, such as process innovations, product and implementation of management strategies, prioritise actions to achieve effective benefits in less time, report environmental performance to regulatory bodies, clients and society, demonstrate the improvements in environmental performance to stakeholders and increased environmental awareness of internal customers, among others (Semarnat, 2014; Medel-Gonzales & García, 2012).

An efficient E.P.I. must have the following characteristics: offer a complete panorama of environmental conditions and pressures; be simple, easy to infer and able to show trends; respond to changes in the environment; provide a basis for comparisons; be applicable at the national or regional level, as the case may be; be theoretically and scientifically well-founded; be able to relate to economic and/or development models, as well as to information systems; be available at a reasonable cost; be well-known, with recognised quality; be easily updated at regular intervals with reliable procedures (OECD, 2003).

5.1.2.2 Environmental Performance Indicators Classification

Due to the large number of environmental performance indicators, there have been several attempts to categorise them. The proposed classifications vary depending on whether the indicator comes from academia, the government sector or other organisations. The classification approach varies according to the author, but there are general trends, such as those proposed by Yale University, the Global Reporting Initiative, the O.E.C.D. and ISO 14031: 99.

Some works such as that of Herva et al. (2011) continue with similar classifications. They suggest four types of E.P.I.: energy material flow indicators, territorial dimension indicators, L.C.A. indicators and indicators based on environmental risk assessment.

Petrovic et al. (2012) has classified E.P.I.s into management condition indicators (M.C.G.) and operational condition indicators (I.C.O.). The former focus on management efforts aimed at facilitating the infrastructure for environmental management. Some of them are environmental objectives have and target programs, personal training and incentive systems to reduce environmental impact, audit on-site inspections, supervision and relations with the community. I.C.O. is related to the operation of the organisation. These include activities such as reuse, emissions reduction and recycling of critical materials, fuel consumption or the use of energy (Petrovic et al., 2012).

On the other hand, Jasch (2000) classifies environmental performance indicators according to the form in which the values are expressed and according to their use and application. These categories are absolute indicators, relative indicators, indexed indicators, global representations and weighted evaluations. Although this standard is based on environmental management system implementation, those activities are linked to stages across the product life. Furthermore, this classification is composed of three type of indicators: operational performance indicators, management performance indicators and environmental condition indicators.

Table 5.1 provides information about how the industry behaves with all the stakeholders. Furthermore, it provides information about how each product and service impact the environment. This impact shows a reactive perspective, such as soil impact and flora decreasing, but also a pro-active perspective, such as design and

TABLE 5.1
Classification of Environmental Performance Indicators

Indicator Type	Dimension	Activities
Operational performance indicator	**Income indicator**	Materials
		Energy
		Services supporting organisation
	Physical facilities and equipment indicator	Design
		Installation
		Operation
		Maintenance
		Land use
		Transportation
	Outcome indicator (provided by the organisation)	Products
		Services
		Wastes
		Emissions
Management performance indicator	System indicator	Policies and programmes implementation
		Compliance
		Financial performance
		Employee participation
	Functional areas indicator	Administration and planning
		Purchases and investments
		Health and security
		Relationships with the community
Environmental condition indicators	Environmental condition Indicators	Emissions
		Water consumption
		Soil impact
		Critical raw material consumption
	Biotic and tropospheric indicators	Flora decreasing
		Wildlife impact
		Aesthetics, heritage and culture

transportation. Although they are indicators of extensive coverage, they can be used to focus the attention of the organisation on the new management of environmental aspects associated with Industry 4.0.

According to Bonilla et al. (2010), although there is literature related to the opportunities, challenges and barriers faced by Industry 4.0 with respect to environmental sustainability, there is no research on the measures and the viability of the decrease of that impact. Although the digitisation of industrial production is beginning, it is still not known if it will promote or hinder the transformations that will affect environmental sustainability. Given that this technology has not been explored from a sustainability perspective due to its novelty and the impossibility of measuring the degrees of implementation in each country or region, although at first sight it is promising, its long-term effects are still unclear.

5.1.3   Colombian Case Study

5.1.3.1   Pressures Facing the Colombian Manufacturing Sector at National and Regional Level

To achieve a balance between cost and sustainability, organisations have implemented policies aimed at correct environmental management. In the case of Colombian Industry 4.0, the way organisations implement these environmental management policies and how they complete their practices throughout the production cycle are unknown. This is despite it being a subject of highly strategic importance for the competitive future of the region.

Changes in the industrial and economic landscape are tempered by phenomena that obey a high complexity in production, given that there is no exact way to describe all products and processes. While this manifests itself as less predictable, highly flexible and sensitive to small externalities, it is precisely this characteristic that makes traditional environmental measurement systems appropriate (Fernandez, 2017).

According to Figure 5.1, some national pressures are the environmental panorama, the consumer pressure and international regulations, as well as the impact generated by manufacturing on the environment. On the other hand, the trends of this sector in the face of new environmental policies are presented, in order to highlight the scarce knowledge that exists about indicators as a source of measurement of all the economic, environmental and social benefits.

According to Figure 5.1, pressures facing Industry 4.0 in Colombia are classified as internal, external and market. Some of the strongest pressures in Colombia are related to the lack of new normative policies with Industry 4.0, as well as the S.M.E. industries that are the main type of enterprises in Colombia.

Among the internal pressures facing Colombia's Industry 4.0 are the lack of trained personnel for this type of system, the lack of access to its own technology and import restrictions, which represents a greater risk on investment, as well as demands for greater capacity and flexibility in systems that are personalised.

In Colombia, all the obstacles and market failures that affect competitiveness must be resolved. But above all, they must design and implement a strategy that leads the large productive groups of the world to consider that it makes sense to insert local productive facilities into these large world flows.

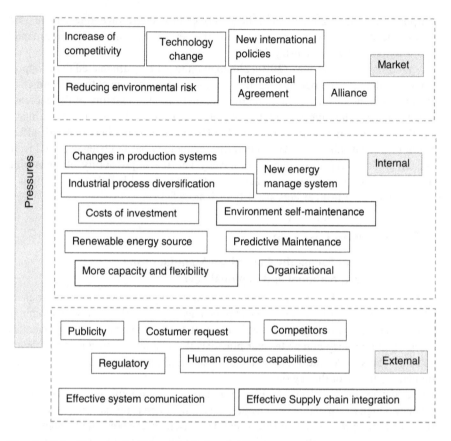

FIGURE 5.1   Pressures in Colombian Industry 4.0.

5.1.3.2   Industry 4.0 Sector in Colombia

Manufacturing is an important sector for the economy. According to a study conducted by Mitchell (2011) for the National Planning Department, in Colombia the industry represents 14.4% of GDP and contributes 12.7% of employment. In the current economic circumstances, and with the signing of the new Free Trade Agreements, the Colombian industry is facing new problems such as environmental concerns by customers and stricter international policies on raw material use and disposal. To confront them, companies must take appropriate actions to achieve better performance and ensure their long-term survival.

In general, companies in the region must confront the new paradigms of competitiveness, where the environment is positioned as a competitive priority. It is for these reasons that measures on manufacturing green practices can generate the development of strategies which through the various activities proposed, such as good relations with suppliers, substitution of materials, integral use of the six Rs, among other activities (Gavronski et al., 2011), along the supply chain, has enabled this productive region to be powerful in other areas (Diabat & Govindan, 2011).

Colombian industry is still in the process of developing. However, many organisations have realised the significance of optimising their processes throughout new technologies such as I.o.T., C.A.D. and C.A.M., among others and have adopted more flexible positions for its application.

However, it is recognised that among the obstacles for the implementation of Industry 4.0 are: the lack of broadband connectivity, the training of highly qualified personnel, the culture of change and return on investment (R.O.I.). Likewise, given the current legislation, the investment risks are high at the technological level and the qualification level of the personnel and the organisational culture limit the incorporation of Industry 4.0 in business contexts.

Although the review of the literature has accounted for multiple developments in the subject matter referred to in Industry 4.0 and its contribution to the development of the economy and for the industry in general, there was a lack of methodological proposals that contribute to the generation of implementation strategies for Industry 4.0 in the local, regional and national context in countries with so many barriers as Colombia.

In this sense, a real policy from the government and all the stakeholders' commitment on Industry 4.0 is required, especially in Colombia, where it is mandatory to promote the incorporation of technology in search of competitive advantages, development and innovation, from the S.M.E.s to big industry. This innovation starts with small pilot exercises and the generation of macro-strategies that enable the country's access to international development.

5.1.4   CONCLUSIONS AND RECOMMENDATIONS

Industry 4.0 has among its conditions the lower consumption of resources, and energy efficiency. However, there is not enough information to say that there is an improvement in the environmental performance of companies that have decided to implement this type of technology or tools, such as embedded or modular systems.

Conventional systems for measuring environmental performance in companies cannot be applied to Industry 4.0, since the limits of the system are blurred, due to the integration of physical and virtual systems and the connection that must exist between all the links of the chain. Another reason that hinders the measurement is the large number of combinations of materials and features in the products generated with this technology.

The environmental performance indicators are strategies to determine the impact of the corrective and proactive actions of the companies, against the attributes of machinery, material, labour, energy and soft technology strategies. Due to their flexibility, they are a potential way to measure the environmental performance of Industry 4.0.

Although the environmental performance indicators are a sample of how Industry 4.0 can be measured, it is necessary to create reliable metrics and comparison values to determine its true impact.

Colombia, due to its historically agricultural economy and the great majority of S.M.E.s that exist, does not have a great development in Industry 4.0. Currently, the government and companies face external challenges, such as new customer requirements, access to new technologies and sensors, internet capacity and communication systems. In addition, the need for clean energy, capacity and flexibility are complex in industrial ecosystems of the artisanal type.

It is necessary to establish clear limits to compare the impact of Industry 4.0 against traditional systems, as there are hidden impacts. For example, although it is true that Industry 4.0 can generate less material consumption, on the other hand, it requires technologies that, in their construction, use scarce critical materials such as rare earth elements or other elements, and their extraction generates a much higher environmental impact.

The Colombian government must establish clear policies and look to the future when thinking about environmental assessment guidelines different from the current ones, since the current environmental performance measures are reactive and only focus on the output of the processes, the amount of water consumed and the emissions generated, and there is no holistic view of the impact.

REFERENCES

Alexopoulos, Ilias, Kounetas, Kostas, & Tzelepis, Dimitris. (2018). Environmental and financial performance. Is there a win-win or a win-loss situation? Evidence from the Greek manufacturing. *Journal of Cleaner Production*, 197(1), 1275–1283, ISSN 0959-6526. doi:10.1016/j.jclepro.2018.06.302.

Atkinson, G. D., Dubourg, R., Hamilton, K., Munasignhe, M., Pearce, D.W., & Young, C. (1997). *Measuring Sustainable Development: Macroeconomics and the Environment.* Edward Elgar, Cheltenham.

Bonilla, S., Almeida, C., Giannetti, B., & Huisingh, D. (2010). The roles of cleaner production in the sustainable development of modern societies: An introduction to this special issue. *Journal of Cleaner Production*, 18, 1e5.

Cámara de comercio de Manizales, Recovery January 09, 2018. Available in http://www.ccmpc. org.co/.

DANE, Departamento Nacional de Planeación, Recovery January 09, 2018, Available in https://www.dane.gov.co/.

de Sousa Jabbour, Ana Beatriz Lopes, Jabbour, Charbel Jose Chiappetta, Godinho Filho, Cyril Foropon, Moacir. (2018). When titans meet – Can industry 4.0 revolutionise the environmentally-sustainable manufacturing wave? The role of critical success factors.

Diabat, A., & Govindan, K. (2011). An analysis of the drivers affecting the implementation of green supply chain management. *Resources, Conservation and Recycling*, 55(6), 659–667. doi:10.1016/j.resconrec.2010.12.002.

Diana, Gabriel Cepollaro, Jabbour, Charbel José Chiappetta, de Sousa Jabbour, Ana Beatriz Lopes, & Kannan, Devika. (2017). Putting environmental technologies into the mainstream: Adoption of environmental technologies by medium-sized manufacturing firms in Brazil. *Journal of Cleaner Production*, 142, Part 4, 4011–4018, ISSN 0959-6526. doi:10.1016/j.jclepro.2016.10.054.

Eaganl, P. D. (1997). Development of a facility-based performance indicator environmental to sustainable development. *Journal of Cleaner Production*, 5(4), 269–278. https://doi. org/10.1016/s0959-6526(97)00044-9.

Eltayeb, T. K., Zailani, S., & Ramayah, T. (2011). Green supply chain initiatives among certified companies in Malaysia and environmental sustainability: Investigating the outcomes. *Resources, Conservation and Recycling*, 55(5), 495–506. doi:10.1016/j. resconrec.2010.09.003.

Epstein, M. J. (2008). *Making Sustainability Work: Best Practices in Managing and Measuring Corporate Social, Environmental, and Economic Impacts.* Greenleaf Publishing Limited, Sheffield.

Fahimnia, Behnam, Jabbarzadeh, Armin, & Sarkis, Joseph. (2018). Greening versus resilience: A supply chain design perspective. *Transportation Research Part E: Logistics and Transportation Review*, 119, 129–148. ISSN 1366-5545, doi:10.1016/j.tre.2018.09.005.

Fernandez, L. (2017). Actas de ingeniería. Industry 4.0: A literature review. La industria 4.0 una revisión de la literatura, 222–227.

Fu Jia, Yu Gong, & Brown, Steve. (2018). Multi-tier sustainable supply chain management: The role of supply chain leadership. *International Journal of Production Economics*, ISSN 0925-5273. doi:10.1016/j.ijpe.2018.07.022.

Gavronski, I., Klassen, R. D., Vachon, S., & Nascimento, L. F. M. do. (2011). A resource-based view of green supply management. *Transportation Research Part E: Logistics and Transportation Review*, 47(6), 872–885. doi:10.1016/j.tre.2011.05.018.

Giljum, S., Burger, E., Hinterberger, F., Lutter, S., & Bruckner, M., (2011). A comprehensive set of resource use indicators from the micro to the macro level. Resources, Conservation & Recycling, 55(1), 300–308.

Hart, S. L., & Milstein, M. B. (2003). Creating sustainable value. *The Academy of Management Executive*, 17(2), 56–67.

Henri, J.-F., & Journeault, M. (2008). Environmental performance indicators: An empirical study of Canadian manufacturing firms. *Journal of Environmental Management*, 87(1), 165–176. doi:10.1016/j.jenvman.2007.01.009.

Heras-Saizarbitoria, I., Landín, G. A., & Molina-Azorín, J. F. (2011). Do drivers matter for the benefits of ISO 14001? *International Journal of Operations & Production Management*, 31(2), 192–216. doi:10.1108/01443571111104764.

Herva, M., Franco, A., Carrasco, E. F., & Roca, E. (2011). Review of corporate environmental indicators. *Journal of Cleaner Production*, 19(15), 1687–1699. doi:10.1016/j.jclepro.2011.05.019.

Hervani, A. A., Helms, M. M., & Sarkis, J. (2005). Performance measurement for green supply chain management. *Benchmarking: An International Journal*, 12(4), 330–353. doi:10.1108/14635770510609015.

Hourneaux, F., Hrdlicka, H. A., Gomes, C. M., & Kruglianskas, I. (2014). The use of environmental performance indicators and size effect: A study of industrial companies. *Ecological Indicators*, 36, 205–212.

Jasch, C. (2000). Environmental performance evaluation and indicators. *Journal of Cleaner Production*, 8(1), 79–88. doi:10.1016/S0959-6526(99)00235-8.

Koberg, Steban, & Longoni, Annachiara (2019). A systematic review of sustainable supply chain management in global supply chains. *Journal of Cleaner Production*, 207, 1084–1098. doi:10.1016/j.jclepro.2018.10.033.

Lee, G. K. L., & Chan, E. H. W. (2009). Indicators for evaluating environmental performance of the Hong Kong urban renewal projects. *Facilities*, 27(13/14), 515–530. doi:10.1108/02632770910996351.

Liu, Yang, Zhu, Qinghua, & Seuring, Stefan. (2017). Linking capabilities to green operations strategies: The moderating role of corporate environmental proactivity. *International Journal of Production Economics*, 187, 182–195. doi:10.1016/j.ijpe.2017.03.007.

Medel-Gonzalez, F., & García, L. F. (2012). *Procedimiento para la evaluación del desempeño ambiental. Aplicación en centrales eléctricas de la UEB de generación distribuidas de Villa Clara*. Santa Clara, Cuba.

Mitchell, D. (2011). *Balance del sector industrial*. PND, Bogotá, Colombia.

Müller, J.M. Kiel, D., & Voigt, K.-I. (2018). What drives the implementation of industry 4.0? The role of opportunities and challenges in the context of sustainability. *Sustainability*, 10, 247.

Neely, A., Adams, C., & Kennerley, M. (2002). *The Performance Prism*. Pearson Education Limited, Edinburgh.

OECD. (2003). Annual report. Organization for Economic Cooperation and Development.

Petrovic, N., Slovic, D., & Cirovic, M. (2012). Environmental performance indicators as guidelines towards sustainability. *Management - Journal for Theory and Practice of Management*, 17(64), 5–14. doi:10.7595/management.fon.2012.0026.

Raihanian Mashhadi, A., & Behdad, S. (2017). Environmental impact assessment of the heterogeneity in consumers' usage behavior: An agent-based modeling approach. *Journal of Industrial Ecology*, 22, 706–719. doi:10.1111/jiec.12622.

Rodríguez-Pose, Andrés, & Wilkie, Callum. (2018). Strategies of gain and strategies of waste: What determines the success of development intervention? *Progress in Planning*, ISSN 0305-9006. doi:10.1016/j.progress.2018.07.001.

Technological forecasting and social change, 132, 2018, 18–25, ISSN 0040-1625. doi:10.1016/j.techfore.2018.01.017.

Tseng, Ming-Lang, Islam, Md Shamimul, Karia, Noorliza, Fauzi, Firdaus Ahmad, & Afrin, Samina. (2019). A literature review on green supply chain management: Trends and future challenges. *Resources, Conservation and Recycling*, 141, 145–162. doi:10.1016/j.resconrec.2018.10.009.

World Economic Forum. (2017). *Informe de riesgos mundiales 2018*. edición 13. Marsh & McLennan Companies.

Zhu, Q., Sarkis, J., & Lai, K. (2008). Green supply chain management implications for "closing the loop." *Transportation Research Part E: Logistics and Transportation Review*, 4(1), 1–18.

5.2   ERP SYSTEMS AND SCM IN INDUSTRY 4.0

E. Manavalan, K. Jayakrishna, E. Vengata Raghavan, and S. Uma Mageswari

5.2.1   INTRODUCTION

One of the challenges for organisations is adapting to new technologies so that product development and innovation happen subsequently to withstand the competition (Ye & Wang, 2013; Czajkiewicz, 2008; Brettel et al., 2014). Organisations have started bringing in innovative solutions to stay ahead of the competition and transform customer expectations into reality with the help of digital technology (Cisneros-Cabrera et al., 2017). Industry 4.0 leverages digital technologies so that interaction happens between components, machines and humans to bring personalised products within the mass manufacturing process (Kagermann et al., 2013). The intelligence of a smart production system is such that it self-diagnoses problems and provides the recommended actions to operations – to be precise, it acts as a virtual assistant (Kang et al., 2016). Industry 4.0 helps the organisation to make suppliers, manufacturers, customers interact seamlessly with the digital world and receive relevant information (Liao et al., 2017).

One of the key expectations of organisations in Industry 4.0 era is to get intelligent information with appropriate actions. The leap in technology allows organisations to achieve actionable reporting made possible with technologies such as Big Data, artificial intelligence and the internet of things. The primary objective of this chapter is to explore the potential opportunities available in S.C.M. with the help of E.R.P. systems and Industry 4.0. Further, it discusses how Industry 4.0 helps the environment and safety of systems and people across the product value chain in a cradle-to-cradle economy.

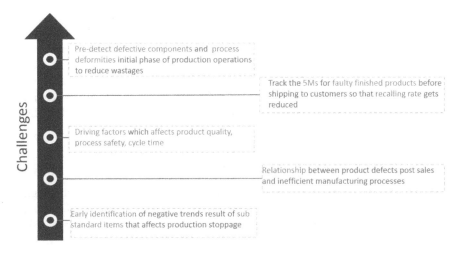

Pre-detect defective components and process deformities initial phase of production operations to reduce wastages

Track the 5Ms for faulty finished products before shipping to customers so that recalling rate gets reduced

Driving factors which affects product quality, process safety, cycle time

Relationship between product defects post sales and inefficient manufacturing processes

Early identification of negative trends result of sub standard items that affects production stoppage

Challenges

FIGURE 5.2 Typical challenges in current E.R.P. systems and S.C.M.

5.2.2 CHALLENGES IN THE SUPPLY CHAIN

The supply chain sector is facing significant challenges, especially in adhering to regulations enforced by local bodies and environmental concerns, and interestingly, every year, the intensity keeps increasing (Coyle et al., 2015). From a wider perspective, every organisation's objective is to increase its revenue and profitability, while at the same time meeting customer expectations. In addition, organisations want to drive optimised manufacturing process to improve their operational efficiency and process safety aspects. To improve profitability, global leaders are striving to reduce production and supply chain costs without compromising on environmental factors (Um et al., 2017).

Some of the challenges faced by the supply chain environment from a product quality and process safety perspective are illustrated in Figure 5.2 (Vanderroost et al., 2017). The stakeholders are finding it difficult to get the operational parameters in real time. Indeed, neither the production performance report nor the operational report give the required details to the stakeholders.

Many countries have come up with stringent regulations to use specific materials and processes to dispose of wastages so that the safety of workers and the environment is protected (Ruiz-Benitez, et al., 2017). The objective is to provide a healthier environment and proper care to consumers. This healthy initiative leads the supply chain organisation to focus on the 6R concepts of remanufacture, reuse, recover, recycle, redesign and reduce (Yan & Feng, 2014).

Manufacturing companies also need to get product compliance across the supply chain from raw materials to the end-item to ensure the sustainability of the product (Stock & Seliger, 2016). The concept of a sustainable supply chain is growing to improve the value chain of the product (Waibel et al., 2017). It envisions extending the product's lifetime. Moreover, supply chain organisations also started looking into different dimensions of sustainability such as environmental, social and economic factors (Husted & Sousa-Filho, 2017). From the environment perspective, manufacturers should spend additional funds, if needed, to dispose of the pollutants and reduce the impact on the environment (Dangelico & Pujari, 2010). The next factor is social sustainability where the organisation faces concerns about providing safe working environments and secure jobs to workers. A healthy atmosphere should be provided to make human and technology more efficient (Missimer et al., 2017). From an economic sustainability standpoint, there is an increase in awareness of reducing energy and water usage. Organisations should not only focus on achieving profits, but must optimise the use of resources to the greatest extent possible (Ding et al., 2017).

5.2.3   Phases of Product Value Chain

The supply chain sector has evolved for many years with the help of technological developments. These technological developments influence industries to transform their organisations to meet Industry 4.0 requirements. This specific section describes phases of the product value chain where the product value increases in every phase. It starts with extracting the raw material and moving it to a warehouse for further processing. The next function in the product development is the design phase where the leading manufacturers look for something unique in the product and develop the product design (Schöggl et al., 2017).

On successful creation of a prototype, the real product is manufactured where organisations focus on core competencies and push themselves to optimise the manufacturing process for mass production (Hedberg Jr, 2017). After all the processing is complete, the finished product is sent for a quality check where different types of product testing are performed. The inspected products are picked, packed and shipped to distribution centres across the globe by logistics agencies (Pauliuk et al., 2017). The sales department establishes the product pricing and markets it to the customers. The sales and marketing management constantly get feedback from customers and are instrumental in improving product development. Later, the product is used by customers in real time where organisations can prove their quality and trust in the product (Abramovici et al., 2017). Then, the product faces the end of life phase, where it is no longer in a usable condition (Helo et al., 2017; Lee et al., 2017).

In the Industry 4.0 era, organisations have started looking to extend the life of the product by adopting digital technologies to reduce wastage and make the work more environmentally friendly (Siemieniuch & Sinclair, 2015). The 6R concept is widely adopted by organisations; 6R stands for remanufacture, reuse, recover, recycle, redesign and reduce (Genovese et al., 2017; Li et al., 2017). The organisation follows any one or many of the 6R concepts to extend the life of the product, which is vital for customer perception and the performance of the product. This is often called a reverse supply chain (R.S.C.), which takes care of returns of the product to save the environment (Sasikumar & Kannan, 2008a, 2008b).

5.2.4   Capitalising on Industry 4.0 Technologies in Supply Chain

Technological advancements in E.R.P. systems help supply chains to drive the economic growth of the organisation (Ferrera et al., 2017). In recent years, there has been a massive transformation in information systems, cloud computing and business analytics that help manufacturing to shift gears and pave the way for a major transformation in this sector (Zhang et al., 2014). Digitalisation provides quick outcomes in the fourth industrial revolution and provides better manufacturing performance (Müller et al., 2017).

5.2.4.1 Influence of Industry 4.0 in Supply Chain

Technology is the key to help organisations to cope up with customer expectations and global competition to meet the dynamic demands. The following section discusses the technology that can make supply chain systems more intelligent and make the fourth industrial revolution more successful. Industry 4.0 offers smart production systems, and hence smart factories produce smart products (Longo et al., 2017). Fundamentally, smart factories are being built with emerging technologies as illustrated in Figure 5.3. This platform helps supply chain organisations to expand the product footprint and meet Industry 4.0 requirements (Rüßmann et al., 2015).

5.2.4.1.1 Autonomous Robots

Automatically self-adjusting, flexible and cooperative machines will be used as per the product, such as R.F.I.D. tags in the assembly line, which will interact with one another, work safely side-by-side with humans and learn from them.

5.2.4.1.2 Simulation

The process will leverage real-time data with the help of Big Data from production touchpoints and the computing power of the cloud to mirror the physical world in a virtual model, which can include machines, products and humans to help set up machines efficiently for a new product line and will enable rapid testing and therefore require more innovation.

5.2.4.1.3 Big Data Analytics

The collection and comprehensive evaluation of data are to identify patterns for predictive maintenance and also for enabling more targeted innovation, marketing and decision-making.

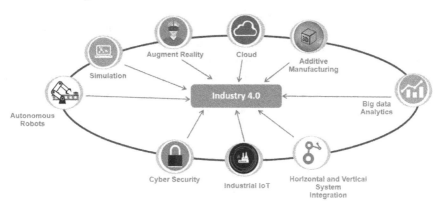

FIGURE 5.3 Industry 4.0 technologies influence S.C.M.

5.2.4.1.4   Industrial I.o.T.

Embedded computing sensors in unfinished products will enable organisations to communicate with the other devices in a production line as necessary, which will facilitate more decentralised decision-making and real-time responses.

5.2.4.1.5   Cybersecurity

Increased connectivity among devices and use of standard communication protocols in Industry 4.0 will increase the need to protect critical industrial systems and manufacturing lines from cybersecurity threats.

5.2.4.1.6   Cloud

Industry 4.0 will require more data sharing across functions, sites and company boundaries, which will be enabled by cloud services and their real-time responses and scalability.

5.2.4.1.7   Additive Manufacturing

Additive manufacturing methods such as 3D printing will be used to add construction advantages to produce small batches of customised products.

5.2.4.1.8   Augmented Reality

Augmented reality will provide workers with real-time information to improve decision-making and productivity during repair and training activities.

5.2.4.1.9   Horizontal and Vertical Integration

The collaboration among enterprise departments and functions and also between partners across value chains will be the result of the enterprise view of data and network systems that enable truly automated value chains.

5.2.5   THE DIGITAL TRANSFORMATION OF SUPPLY CHAIN IN INDUSTRY 4.0

Industry 4.0 makes a product value chain to achieve incremental growth and cost reduction. Industry 4.0 is an amalgamation of digital technologies that make components, machines and humans interact in the production environment (Schweer & Sahl, 2017). This section describes the Industry 4.0 applications across different stages of the product value chain with the focus on reducing the environmental footprint and enhancing process safety (Riel et al., 2017). The chapter also explores the Industry 4.0 offerings across the supply chain, particularly in the areas of operations and process control. The discussed use cases are predominantly from environmental and safety control perspectives, which drive environmental energy management and safety management in the supply chain.

5.2.5.1   Raw Materials and Raw Materials Processing

Advanced materials, which have lower weight and improved energy use are to be used in production so that the carbon footprint can be reduced (Tu et al., 2017). From a health and safety management perspective, as illustrated in Figure 5.4, drones can

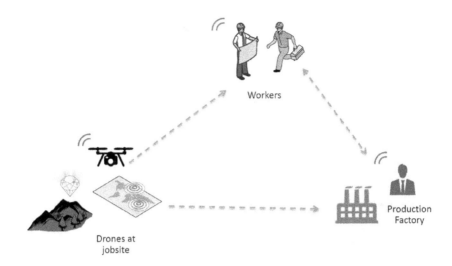

FIGURE 5.4   Connected production system with drones deployed at job site.

be used for raw material extraction in places where involving humans is unsafe, especially in steel production, mining and chemical oil and gas industries. Industry 4.0
also enables industries to avoid potential shutdowns by planning personalised maintenance programmes and retaining efficient manufacturing processes. It improves
inspections and worker safety and identifies potential risks before they can become
hazardous. Production organisations will truly benefit from the use of unmanned
aerial vehicles when it comes to overall safety of workers (Qiuping et al., 2011).

5.2.5.2   Design
Product design with additive manufacturing allows product designers to focus on
performance, which affects the environment and product safety, as this technology eliminates the traditional barriers of design for assembly or manufacturing
(Kochan & Miksche, 2017). Additive manufacturing technology helps manufacturers to directly manufacture components from digital drawings. Thus, it significantly
reduces product development cost and lead-time in research and development. In
other words, less wastage allows manufactures to make environmentally friendly
products. More importantly, consumers get the customised items and manufacturers
benefit from mass production.

As illustrated in Figure 5.5, car manufacturers started using 3D printing technology to showcase the vehicle design models of new 3D products. Additive manufacturing allows organisations to make the product from a more economical perspective,
and it also ensures that products reach the customers faster.

Virtual reality is another technology which helps organisations to reduce the
product development time. Products are simulated in virtual reality that allows the
design team to walk through the product design and identify the design issues virtually without even prototyping it. Designers can view the end-to-end impact of a
design change without building the prototype. Major manufacturers started using

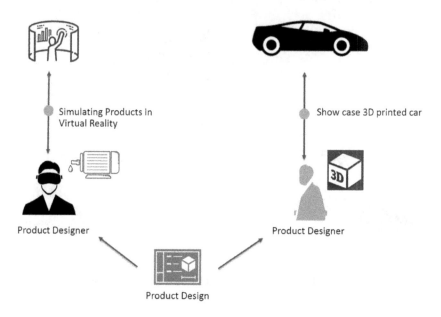

FIGURE 5.5 Various applications of virtual reality and additive manufacturing.

virtual reality as a virtual assistant. It is possible now for photorealistic products to be created using laser scanners (Paelke, 2014).

5.2.5.3 Manufacturing

A smart system consists of four key components: smart products, efficient processes, informed people and intelligent infrastructure. These components share real-time information about parts, machines and equipment and interact with humans to ensure a seamless manufacturing process (Li et al., 2017). C.P.S. allows machines to self-organise and recommends appropriate actions to stakeholders (Bocciarelli et al., 2017). People across different areas get real-time information to meet the quality service and also to ensure the safety of the assets.

The safety of the processes is ensured by monitoring the key indicators such as potential maintenance issues like temperature fluctuations, excessive vibrations and abnormal patterns remotely in real time. In addition, the smart production system monitors operational inputs such as pressure and position continuously and communicates these inputs to stakeholders so that the necessary actions will be taken. Industry 4.0 helps to make the entire production visible across the enterprise from remote locations. Suppliers get visibility into inventory status on the shop floor and supply items just in time to avoid inventory-carrying costs as well as stock-out costs.

5.2.5.4 Distribution

With Industry 4.0, a warehouse is not just for storing inventory; it proactively boosts the efficiency of the supply chain. A smart warehouse consists of wearable I.o.T. devices, smart equipment and cloud technology, transforming distribution management. The advances in technology enable warehouse operators to locate the exact

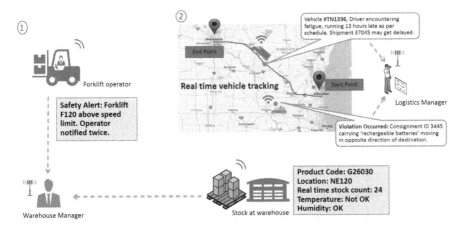

FIGURE 5.6 Smart warehouse and real time vehicle tracking with IoT devices.

storage location of material and monitor its condition from anywhere in the warehouse. Wearable I.o.T. devices improve safety of workers as they reduce physical handling of material, especially when it comes to hazardous and heavy-duty items. Use of I.o.T. in distribution management is growing these days and it has a ripple effect throughout the supply chain. Organisations have benefited by investing in I.o.T., which can improve speed and shipment accuracy and reduce manual intervention. As illustrated in Figure 5.6, it is evident from the mentioned use cases that a smart warehouse definitely improves operational and system efficiency with I.o.T. devices.

The warehouse manager is notified with real-time information from the warehouse, which incidentally protects stock safety as well as worker safety. Traditionally, logistics management is a challenging task for a logistics manager, as it functions around the clock for almost the entire year. The manufacturer needs to ensure that stocks are available as per demand.

With the help of I.o.T., a global positioning system allows warehouse managers to track the fleet. It also sends users alerts and notifications about the shipment status.

5.2.5.5 Sales

Smart products create new opportunities to transform value propositions and allows up-selling them in different markets. This requires a technology influence to ensure that there are undisrupted sales and works seamlessly to absorb sudden spikes in demand for the products.

Organisations get detailed alternate item analysis from Big Data analytics, when near stock-out happens in a specific sales region. The intelligent system not only alerts the sales manager about insufficient stock, it also allows the user to perform alternate item analysis. As illustrated in Figure 5.7, when the system predicts that the stock outage may likely happen soon, the sales manager is provided with options to get it from a nearby strategic partner with an additional cost or get it from a company-owned warehouse without any additional cost.

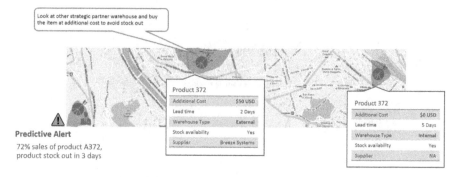

FIGURE 5.7 Redefine sales with Industry 4.0.

Based on the transportation lead-time and customer relationship record, sales managers can take decisions to get it from a strategic partner and absorb the additional cost to avoid stock-out and ensure customer demand is fulfilled. It largely improves the company branding and economic safety factors.

5.2.5.6 Use Phase

With the volume of data that exists in E.R.P., Big Data analytics helps in making better decisions and in product safety. Earlier organisations were using descriptive analytics, which provides insight into the past. It gives answers about what the past data suggests. Recently, organisations started using predictive analytics, which uses forecast models and predicts the future. This provides an answer for what is foreseen and has a probable chance of occurrence. Furthermore, prescriptive analytics is emerging among supply chains that use optimisation algorithms, and it comes up with a simulation of recommended solutions for the possible outcome. This helps on what action is to be taken in the given situation in Industry 4.0.

One of the real-time use cases that is discussed is illustrated in Figure 5.8, which shows the importance of prescriptive analytics in the use phase of a product. I.o.T. helps in tracking the usage of the bicycle and finds any issues such as paint cracks, wheel misalignment, etc. With I.o.T., the E.R.P. system identifies the issue on the operational parameters such as temperature and humidity during painting.

Perspective analytics analyses the possible reasons for paint cracks, looking at various combinations of temperature and humidity. Later it uses optimisation algorithms and simulation techniques and suggests to the production manager the optimum temperature and humidity to avoid paint cracks and ensure the safety of bicycle riders. Through this, better and faster decisions are taken at the production facility. Using adaptive intelligence, the system recommends the next course of action in a timely and effective manner.

5.2.5.7 Extended Life of the Product – Make a Sustainable Impact

Today supply chains are looking to improve their competitiveness with advanced technologies which use I.C.T. to secure sustainable growth. Using innovative technologies along with E.R.P. systems, the industry tries to extend the life of their products by applying any of the 6R concepts. Every stakeholder in the supply chain is responsible for making the world eco-friendly.

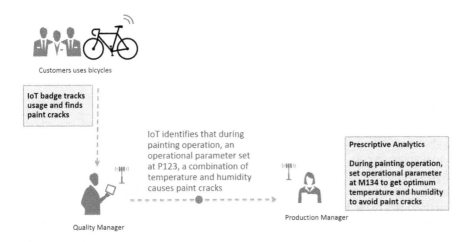

FIGURE 5.8 Prescriptive analytics in use phase of a product.

5.2.5.7.1 I.o.T.-Enabled Technology in Environmental Industry

In this section, use cases of I.o.T.-enabled H.V.A.C. industry and production industry are discussed. As illustrated in Figure 5.9, the first part shows the recycling process enabled by I.o.T. When an item is scrapped during work order processing, the production operator puts the recyclable items into an I.o.T.-embedded bin. A smart device predicts the filling pattern and notifies the operations manager as well as the logistics manager. The logistics manager plans for an optimised pickup schedule well in advance. Further, the operations manager tracks the fleet status and service performance in real time.

The second part of the use case is about the I.o.T.-enabled H.V.A.C. industry where the organisation deals with multiple plants. Practically, it would be difficult to assess the H.V.A.C. unit efficiency manually. In this use case, H.V.A.C. unit reports

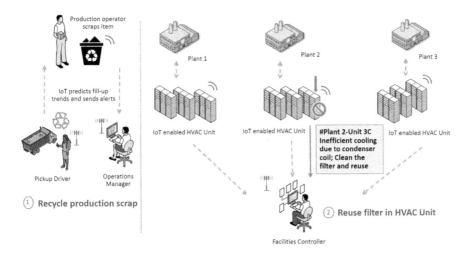

FIGURE 5.9 Recycle and reuse items in extended phase of a product.

cooling inefficiency with the suggestion to clean the filter and reuse it again to improve efficiency. In addition, smart devices, which are in the form of thermostats and environmental condition tracking sensors, assess weather conditions and adjust the temperature dynamically with the help of the H.V.A.C. unit.

5.2.6 Conclusion

The new product innovation accelerates prototyping, improves the ability to capture and respond to customer needs faster, and reduces product development lead-time and innovation costs. Organisations adhering to an eco-friendly environment reduce their carbon footprint and reduce operations and maintenance costs. More importantly, with Industry 4.0 components, the safety of people and processes is monitored continuously in the E.R.P. system.

In this chapter, the challenges in the supply chain are discussed initially. Later, the entire phases of the product value chain across the supply chain are described. Further, how the product life can be extended with the 6Rs, namely, remanufacture, reuse, recover recycle, redesign and reduce, is studied. Next, a comprehensive study on advanced technologies, which act as the building blocks of the fourth industrial revolution, is discussed. Furthermore, the prospects of an E.R.P. system and S.C.M. in the Industry 4.0 environment are presented, with the focus on reducing the environmental footprint and enhancing process safety. Further advancements in Industry 4.0's underlying technologies such as I.o.T., cloud computing, automation and analytics create significant opportunities along the supply chain, which improves overall competitiveness and operational efficiency.

ABBREVIATIONS

I.o.T.	Internet of Things
6Rs	Remanufacture, Reuse, Recover, Recycle, Redesign and Reduce
R.F.I.D.	Radio Frequency Identification
R.S.C.	Reverse Supply Chain
I.C.T.	Information and Communication Technology
H.V.A.C.	Heating, Ventilation and Air Conditioning
C.P.S.	Cyber-Physical System
E.R.P.	Enterprise Resource Planning

REFERENCES

Abramovici, M., Göbel, J. C., & Savarino, P. (2017). Reconfiguration of smart products during their use phase based on virtual product twins. *CIRP Annals, 66*(1), 165–168.
Bocciarelli, P., D'Ambrogio, A., Giglio, A., & Paglia, E. (2017, May). A BPMN extension for modeling Cyber-Physical-Production-Systems in the context of Industry 4.0. In *Networking, Sensing and Control (ICNSC), 2017 IEEE 14th International Conference on* (pp. 599–604). IEEE, Calabria, Italy.
Brettel, M., Friederichsen, N., Keller, M., & Rosenberg, M. (2014). How virtualization, decentralization and network building change the manufacturing landscape: An industry 4.0 perspective. *International Journal of Mechanical, Industrial Science and Engineering, 8*(1), 37–44.

Cisneros-Cabrera, S., Ramzan, A., Sampaio, P., & Mehandjiev, N. (2017, September). Digital marketplaces for industry 4.0: A survey and gap analysis. In *Working Conference on Virtual Enterprises* (pp. 18–27). Springer, Cham.

Coyle, J. J., Thomchick, E. A., & Ruamsook, K. (2015). Environmentally sustainable supply chain management: An evolutionary framework. In *Marketing Dynamism & Sustainability: Things Change, Things Stay the Same* (pp. 365–374). Springer, Cham.

Czajkiewicz, Z. (2008). Direct digital manufacturing-new product development and production technology. *Economics and Organization of Enterprise, 2*(2), 29–37.

Dangelico, R. M., & Pujari, D. (2010). Mainstreaming green product innovation: Why and how companies integrate environmental sustainability. *Journal of Business Ethics, 95*(3), 471–486.

Ding, K., Jiang, P., & Zheng, M. (2017). Environmental and economic sustainability-aware resource service scheduling for industrial product service systems. *Journal of Intelligent Manufacturing, 28*(6), 1303–1316.

Ferrera, E., Rossini, R., Baptista, A. J., Evans, S., Hovest, G. G., Holgado, M., … Silva, E. J. (2017, April). Toward industry 4.0: Efficient and sustainable manufacturing leveraging MAESTRI total efficiency framework. In *International Conference on Sustainable Design and Manufacturing* (pp. 624–633). Springer, Cham.

Genovese, A., Acquaye, A. A., Figueroa, A., & Koh, S. L. (2017). Sustainable supply chain management and the transition towards a circular economy: Evidence and some applications. *Omega, 66*, 344–357.

Hedberg Jr, T. D., Hartman, N. W., Rosche, P., & Fischer, K. (2017). Identified research directions for using manufacturing knowledge earlier in the product life cycle. *International Journal of Production Research, 55*(3), 819–827.

Helo, P., Gunasekaran, A., & Rymaszewska, A. (2017). Value chain effects. In Su, F. (Ed.), *Designing and Managing Industrial Product-Service Systems* (pp. 83–87). Springer International Publishing, Switzerland AG.

Husted, B. W., & de Sousa-Filho, J. M. (2017). The impact of sustainability governance, country stakeholder orientation, and country risk on environmental, social, and governance performance. *Journal of Cleaner Production, 155*, 93–102.

Kagermann, H., Helbig, J., Hellinger, A., & Wahlster, W. (2013). *Recommendations For Implementing the Strategic Initiative INDUSTRIE 4.0: Securing the Future of German Manufacturing Industry; Final Report of the Industrie 4.0 Working Group.* Forschungsunion.

Kang, H. S., Lee, J. Y., Choi, S., Kim, H., Park, J. H., Son, J. Y., Kim, B. H., & Do Noh, S. (2016). Smart manufacturing: Past research, present findings, and future directions. *International Journal of Precision Engineering and Manufacturing-Green Technology, 3*(1), 111–128.

Kochan, D., & Miksche, R. (2017, June). Advanced manufacturing and industrie 4.0 for SME. In *International Conference on Advanced Manufacturing Engineering and Technologies* (pp. 357–364). Springer, Cham.

Lee, J., Suckling, J. R., Lilley, D., & Wilson, G. T. (2017). What is 'value' and how can we capture it from the product value chain? In Matsumoto, M., Masui, K., Fukushige, S., & Kondoh, S. (Eds.), *Sustainability through Innovation in Product Life Cycle Design* (pp. 297–313). Springer, Singapore.

Li, D., Tang, H., Wang, S., & Liu, C. (2017). A big data enabled load-balancing control for smart manufacturing of Industry 4.0. *Cluster Computing, 20*(2), 1855–1864.

Li, W., Wu, H., Jin, M., & Lai, M. (2017). Two-stage remanufacturing decision makings considering product life cycle and consumer perception. *Journal of Cleaner Production, 161*, 581–590.

Liao, Y., Deschamps, F., Loures, E. D. F. R., & Ramos, L. F. P. (2017). Past, present and future of Industry 4.0-a systematic literature review and research agenda proposal. *International Journal of Production Research, 55*(12), 3609–3629.

Longo, F., Nicoletti, L., & Padovano, A. (2017). Smart operators in industry 4.0: A human-centered approach to enhance operators' capabilities and competencies within the new smart factory context. *Computers & Industrial Engineering, 113*, 144–159.

Missimer, M., Robèrt, K. H., & Broman, G. (2017). A strategic approach to social sustainability–Part 1: Exploring the social system. *Journal of Cleaner Production, 140*, 32–41.

Müller, J., Dotzauer, V., & Voigt, K. I. (2017). Industry 4.0 and its impact on reshoring decisions of German manufacturing enterprises. In Bode, C., Bogaschewsky, R., Eßig, M., Lasch, R., & Stölzle, W. (Eds.), *Supply Management Research* (pp. 165–179). Springer Gabler, Wiesbaden.

Paelke, V. (2014, September). Augmented reality in the smart factory: Supporting workers in an industry 4.0. environment. In *Emerging Technology and Factory Automation (ETFA), 2014 IEEE* (pp. 1–4). IEEE, Barcelona, Spain.

Pauliuk, S., Kondo, Y., Nakamura, S., & Nakajima, K. (2017). Regional distribution and losses of end-of-life steel throughout multiple product life cycles—Insights from the global multiregional MaTrace model. *Resources, Conservation and Recycling, 116*, 84–93.

Qiuping, W., Shunbing, Z., & Chunquan, D. (2011). Study on key technologies of Internet of Things perceiving mine. *Procedia Engineering, 26*, 2326–2333.

Riel, A., Kreiner, C., Macher, G., & Messnarz, R. (2017). Integrated design for tackling safety and security challenges of smart products and digital manufacturing. *CIRP Annals, 66*(1), 177–180.

Rüßmann, M., Lorenz, M., Gerbert, P., Waldner, M., Justus, J., Engel, P., & Harnisch, M. (2015). Industry 4.0: The future of productivity and growth in manufacturing industries. Boston Consulting Group, 9.

Ruiz-Benitez, R., López, C., & Real, J. C. (2017). Environmental benefits of lean, green and resilient supply chain management: The case of the aerospace sector. *Journal of Cleaner Production, 167*, 850–862.

Sasikumar, P., & Kannan, G. (2008a, November). Issues in reverse supply chains, part I: End-of-life product recovery and inventory management–an overview. *International Journal of Sustainable Engineering, 1*(3), 154–172.

Sasikumar, P., & Kannan, G. (2008b, December). Issues in reverse supply chains, part II: Reverse distribution issues–an overview. *International Journal of Sustainable Engineering, 1*(4), 234–249.

Schöggl, J. P., Baumgartner, R. J., & Hofer, D. (2017). Improving sustainability performance in early phases of product design: A checklist for sustainable product development tested in the automotive industry. *Journal of Cleaner Production, 140*, 1602–1617.

Schweer, D., & Sahl, J. C. (2017). The Digital transformation of industry–the benefit for Germany. In Abolhassan, F. (Ed.), *The Drivers of Digital Transformation* (pp. 23–31). Springer International Publishing, Switzerland AG.

Siemieniuch, C. E., & Sinclair, M. A. (2015). Global drivers, sustainable manufacturing and systems ergonomics. *Applied Ergonomics, 51*, 104–119.

Stock, T., & Seliger, G. (2016). Opportunities of sustainable manufacturing in industry 4.0. *Procedia CIRP, 40*, 536–541.

Tu, M., Chung, W. H., Chiu, C. K., Chung, W., & Tzeng, Y. (2017, April). A novel IoT-based dynamic carbon footprint approach to reducing uncertainties in carbon footprint assessment of a solar PV supply chain. In *Industrial Engineering and Applications (ICIEA), 2017 4th International Conference on* (pp. 249–254). IEEE.

Um, J., Lyons, A., Lam, H. K., Cheng, T. C. E., & Dominguez-Pery, C. (2017). Product variety management and supply chain performance: A capability perspective on their relationships and competitiveness implications. *International Journal of Production Economics, 187*, 15–26.

Vanderroost, M., Ragaert, P., Verwaeren, J., De Meulenaer, B., De Baets, B., & Devlieghere, F. (2017). The digitization of a food package's life cycle: Existing and emerging computer systems in the pre-logistics phase. *Computers in Industry, 87*, 1–14.

Waibel, M. W., Steenkamp, L. P., Moloko, N., & Oosthuizen, G. A. (2017). Investigating the effects of smart production systems on sustainability elements. *Procedia Manufacturing, 8*, 731–737.

Yan, J., & Feng, C. (2014). Sustainable design-oriented product modularity combined with 6R concept: A case study of rotor laboratory bench. *Clean Technologies and Environmental Policy, 16*(1), 95–109.

Ye, F., & Wang, Z. (2013). Effects of information technology alignment and information sharing on supply chain operational performance. *Computers & Industrial Engineering, 65*(3), 370–377.

Zhang, L., Luo, Y., Tao, F., Li, B. H., Ren, L., Zhang, X., Guo, H., Cheng, Y., Hu, A., & Liu, Y. (2014). Cloud manufacturing: A new manufacturing paradigm. *Enterprise Information Systems, 8*(2), 167–187.

5.3 THE IMPORTANCE OF ADDITIVE MANUFACTURING – FACTORIES OF THE FUTURE

Prateek Saxena and Himanshu Singh

5.3.1 INTRODUCTION

Industry 4.0, the next industrial revolution, involves a robust combination of smart production methods and highly superior information technologies that utilises cyber-physical systems (C.P.S.) in the manufacturing environment. In the modern era of intelligent automation, the use of contemporary manufacturing arts from the perspective of combining innovative information technologies plays a significant position in economic development. Industry 4.0 connects cyber and physical systems in a way that means human beings can get services in a simplified manner, which further enhances the efficiency of benefits in their work. The role of human capabilities is redefined by building smart factories.

The factories of the future are driven by an advanced pragmatic environment that includes the internet of things, cloud computing, Big Data, etc. In addition to this, physical environments include A.M. and autonomous robots (Liu & Xu 2016). In connection with the cyberworld and the real world, multiple computer interfaces or stimulated wireless nexus are essential tools that collect a massive amount of information generated from physical objects. Thus, the collected datasets from products, production line constituents and machines contain a lot of statistical data, which further needs to be exchanged and analysed. The existing data can be of any type, including design records, customer orders, supply details, order cancellation details, stock and logistics-related information. In a nutshell, this huge dataset is termed Big Data. It is essential for an interconnected and well-established manufacturing environment to have mobile internet. In this, data collection plays a very vital role, so there is a need for object tagging and internet-to-object conversation. Highly efficient communication requires real-time data capturing and approachability; in this regard, cloud computing can offer productive computing and storage potential for digitally intensified manufacturing systems. All these kinds of virtual technologies create technical support, assuring the better utilisation of surviving pieces of information for intelligent manufacturing, the factories of the future (Lidong & Guanghui 2016; Saxena et al. 2020).

A.M. is considered one of the essential components of Industry 4.0 because the customer demand or more precisely, customer product, and a substantial part of the smart factories of the future are controlled by the performance and capacity of the actual manufacturing systems. With an increase in customer expectations, highly customisable products as per the demand are required. Industry 4.0 promotes the requirement of mass customisation, and there is a need for highly efficient and precise non-traditional production techniques to be improved. Intricate objects with superior properties, such as the latest materials and shapes, can be fabricated by A.M. It is a proven technology to produce quality parts and is widely used in the aerospace, bio-medical and manufacturing industries (Thompson et al. 2016).

5.3.2 Materials for A.M.

Though a lot of plastic/polymers are available for A.M., there is still ongoing research for identifying more suitable and advanced materials for A.M. (Singh et al., 2017). During its inception phase, plastics were used to produce prototypes, and different plastics were used as feed materials in various additive manufacturing processes, such as the Fused Deposition Method (F.D.M.) technique. Recent advancements in developing more complex parts and shapes using A.M. led to the development of more advanced materials. Depending on the application of the intended product, metals, ceramics and composites are used for A.M. processes. This section gives an overview of the types of materials utilised in many A.M. methods.

5.3.2.1 Polymers

A polymer is defined as a 'conglomeration of various monomers, which are basically repeating structural units'. They can be classified as natural or synthetic polymers. In the A.M. process, Acrylobuta styrene (A.B.S.), nylon, photosensitive resins, elastomers and wax are widely used in F.D.M. and Selective Laser Sintering (S.L.S.) processes. Nylon is usually preferred in the S.L.S. process because the particles can easily be melted by laser heat flux, facilitating the ease of bond formation (Kruth et al., 2007). A.B.S., on the other hand, is preferred for the F.D.M. process. Other polymers suitable for A.M. processes may include elastomeric, wax and starch-based polymers (Banack & Sherman, 2013).

5.3.2.2 Metals

Engineering applications, where mechanical strength is a favourable matter of concern, metals are the most common materials among all the other classes of materials. Metal 3D printing is highly promising and preferred in factories of the future, as they have the potential to produce almost all the parts that are fabricated using polymers as the base material. Metal A.M. thus has emerged as an active field of interest in recent years. Industrial components with complexity can be produced with ease by metal A.M. in titanium, aluminium and stainless steel (Herzog et al.,

2016). The input material is a metallic powder for the 3D printer; however, different combinations of metallic powders can also be utilised for the A.M. process (Körner 2016). Mechanical attributes such as tensile strength and fatigue properties of these components are highly dependent on the microstructure of the resulting A.M. products. Thus, the characteristics related to microstructure, phase composition and heat transfer during the manufacturing process is considered while fabricating the parts (Grigoriev et al., 2017; Tang & Pistorius, 2017). With existing technologies, industrial expectations are sometimes not fulfilled by the manufactured product, leaving room and need for improvement. Favourable cost, speed of the production, high tensile/fatigue/ hardness performance, better surface conditions and a homogeneous microstructure are some key parameters that can be enhanced using advanced manufacturing techniques that can be implemented in factories of the future (Sames et al., 2016).

Manufacturing techniques can broadly be classified into direct and indirect A.M. techniques. Using the indirect A.M. technique, the binder is used to bind metal powders to produce 3D objects. Post-processing of the final product is performed at the last stage of the production. Indirect A.M. techniques involve S.L.S., F.D.M., Stereolithography (S.L.A.) and Laminated Object Manufacturing (L.O.M.) techniques. In the direct method, the final product or final complex geometries are obtained by the direct melting of the metallic powder, and the formation of metallic bonds takes place. Direct A.M. methods include Selective Laser Melting (S.L.M.), Electron Beam Melting (E.B.M.) and Laser Metal Deposition (L.M.D.) techniques. In Industry 4.0, it is expected that metal A.M. will be an essential player for the success of the new industrial revolution. The methods are discussed in detail in the next sections.

i) **Indirect metal A.M. technique.** This section gives the reader a more in-depth overview of the indirect A.M. techniques. Selective Laser Sintering is one of the indirect thermo-physical phenomena. In this technique, the final product is manufactured by using a partial melting of metallic powder (Kruth et al., 2007) or by melting the low melting point adhesive whose function is to bind two metallic particles together (Beaman et al., 2002). The main objective of this technique is to create highly dense metallic components. Thus, for achieving such kinds of highly densified product, specific post-processing techniques, such as elimination of polymer binder, thermic sintering and liquid and metal infiltration techniques are performed. Phenolic polymer, low melting temperature metals such as S.n.S. are some binders preferentially used in S.L.S. Non-melting indirect A.M. techniques include S.L.A. and L.O.M. techniques. In S.L.A., metallic powder is spread and a liquid binder is poured onto its top surface which generates the formation of new adhesive bonding between the molecules. In the S.L.A. technique, U.V. light is used to improve the suspension made by blending small metal powders into a liquid photocurable resin.

ii) **Direct metal A.M. techniques.** In this method, a high-intensity laser or electron beam acts as a heat source. The direct approach incorporates additive manufacturing processes such as S.L.M., E.B.M. or L.M.D./L.E.N.S.

Five main binding mechanisms are generally present during S.L.S. and S.L.M., namely, solid-state sintering, chemical-induced adhesion, liquid-state sintering, incomplete melting and full melting. These binding mechanisms largely influence the processing speed and determine the microstructure of the manufactured product. Partial melting and complete melting are the two standard techniques used in most of the commercial products (Kruth et al., 2007; Kruth et al., 2005).

In S.L.S. and E.B.M., a powder bed of metal is used, and these powders are continuously supplied into the melt pool, which is generated by a high energy laser beam. This results in a completely solid metal part that has almost the identical or even better mechanical features as the bulk metal. Deflecting Galvano mirrors are used for controlling the co-ordinates of the sintering location following imported C.A.D. geometry. In the manufacturing of the product using A.M., the strengthening of powders by sintering techniques play a meaningful role. The quality of newly developed microstructures depends upon the integral coalescence when powders are subjected to high sintering temperature (Malik & Kandasubramanian, 2018).

The mechanical characteristics of the metal elements: Stainless steel, tool steel, nickel alloys, cobalt alloys, titanium alloys and a vast range of cladding alloys have previously been processed successfully with L.E.N.S. and S.L.S. techniques. The microstructure and mechanical characteristics of the Inconel 718 manufactured by electron beam melting were explored by Strondl et al. (2009). The E.B.M. method is used for fabricating H13 steel parts (Cormier et al., 2004). Nickel-titanium shape memory alloys were manufactured by E.B.M. (Otubo & Antunes, 2010). Thin wall components, such as turbine blades, and intricate structures, such as diamond frame structures, have been fabricated by these techniques, as shown in Figure 5.10.

5.3.2.3 Ceramics

Ceramic material is defined as an inorganic, non-metallic generally crystalline oxides, nitrides or carbides of metals such as aluminium, zirconium, carbon, silicon,

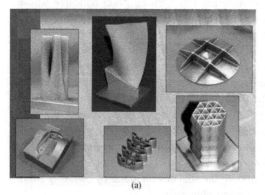

(a)   (b)

FIGURE 5.10 (a) Metal components manufactured by adopting L.E.N.S. (Optomec, n.d.); (b) Fine framework construction for application in the medical field (material: cobalt chrome alloy) made by employing S.L.M. (Concept Laser, n.d.).

etc. The properties of ceramics include temperature-resistance, high brittleness, high hardness, high compressive strength, weak shear and tensile strength. Some examples include earthenware, porcelain and brick. Because of their mechanical properties, these materials are difficult to manufacture. Thus, manufacturing of complex shapes using ceramic material in A.M. is still a challenge. A.M. has successfully demonstrated its unique credentials in the production of ceramic parts using both direct and indirect methods which are discussed in the next section.

   i) **Indirect methods for AM of ceramics.** Ceramics used in industries are mainly nitrides and oxides of metals e.g., silica nitride, aluminium oxide, silica oxide. Advanced ceramics such as lead zirconate and biocompatible ceramics components are fabricated using F.D.M., S.L.A. and S.L.S. techniques. The green mould is first prepared by the addition of organic or inorganic binders. The prepared green mould is compressed by applying high pressure, and then it is put inside the furnace for its sintering at an elevated temperature. At such high temperature, binders are burned out and deification of the final product is achieved. Figure 5.11 shows some parts produced in ceramics using A.M. techniques.

Advanced ceramics are composed of various functional groups such as bismuth titanate, aluminium structures with photonic bandgap properties and piezoelectric actuators (Safari, 2001). F.D.M. is more commonly used for fabrication of structural parts in silica nitride and silica oxide by using a polymer filament loaded with ceramic (Beaman et al., 2002). For the fabrication of highly dense components, a

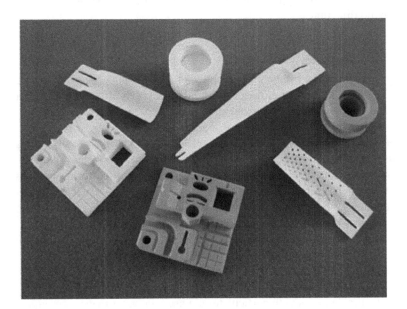

FIGURE 5.11   Alumina and silica ceramic cores manufactured by applying S.L.S. for investment casting of turbine blades and distinct ceramic components (Phenix, 2015).

green mould is prepared by mixing ceramic particles with thermoplastic polymers, and a hot extrusion process is carried out on the product. The object is then desiccated so that removal of the binder can take place, and then put inside the sintering furnace. F.D.M. in the ceramic method is termed a fused deposition of ceramics (F.D.C.). In another study by Sun et al. (2002), 3D printing (3D.P.) was utilised by spreading powder of Ti3SiC2 onto the bed and the binder being sprayed on it. Unique electrical and mechanical characteristics were thus obtained during cold isostatic pressing and sintering at high temperature.

ii) **Direct methods for A.M. of ceramics.** Ceramics have a very high melting point (M.P.), such as for Al2O3 (M.P. >20273 K) and for silica oxide (M.P. >19273 K), and therefore melting is a very challenging task using A.M. processes, such as in laser-based A.M. process. In addition to this, ceramics have substantial thermal gradients, thermal stresses and residual stresses correlated with its melting and solidification. It is, however, possible to manufacture solid ceramic components using A.M. By melting powders of zirconia and alumina using S.L.M., a highly dense ceramic product is produced. At first, the temperature of ceramic powder is increased to 1873 K by placing it in the furnace, thereby reducing the stresses caused by the temperature difference, and producing excellent, crack-free products without some further post-processing. Direct laser melting of ceramic powder is performed using L.E.N.S.E., which results in very dense, net-shaped alumina parts (Balla et al., 2008). The resulting ceramic product displays anisotropy in its mechanical characteristics, having a very high compressive strength value with columnar grains along it, regular to its build direction.

5.3.2.4 Composites

Composites are used for enhancing the physical and chemical properties of the materials. These are natural or designed materials composed of two, or more than two, materials having different physical and chemical properties. The constituent particles exhibit distinct microscopic or macroscopic scale and the properties of the composite are enhanced (Anand & Saxena, 2016). When the constituent particles are dispersed in a uniform phase then, then the resulting product is called a homogeneous compound or uniform composite. When it is non-uniform, it results in an inhomogeneous compound; an example of such kind of materials are functionally graded materials. In these materials, there is a gradual variation of composition over volume, which results in a change in behaviour of the composite materials.

In the perspective of A.M., uniform composites are manufactured by mixing different kinds of materials in the desired proportions. For S.L.M., S.L.S. and 3D.P., a mixed powder bed is created, for F.D.M., a fixed material filament is used. Similarly, in L.O.M. a composite laminate structure is used, and for S.T.L., a solution of liquid photocurable resins with particulates are used. Polymer matrix, ceramic matrix, metal-matrix and fibre- and particulate-reinforced composites are highly used composite materials in A.M. (Kumar & Kruth, 2010). Fibre-reinforced composites are one of the most important composite material used in industrial applications (Saxena et al., 2016). Fibre-based composites can be fabricated using F.D.M. and L.O.M. The main challenge for fabrication of fibre-based composites is that it is very difficult to make a smooth layer of powder fibre mixture.

5.3.3 SMART MATERIALS FOR INDUSTRY 4.0

Smart materials can change their shape and material properties depending upon the change in surrounding conditions. The inclusion of highly advanced smart materials in Industry 4.0 provides the possibility of altering features, such as the configuration of the printed structure, to fabricate practical geometries of the soft robotic system, self-evolving fabrications and controlled regular folding applications. Generally, shape memory alloys (S.M.A.) and shape memory polymers (S.M.P.) are used as smart materials for 4D printing (Khoo et al., 2015).

The super-elasticity and thermal shape restoration features of the shape memory alloys are among the excellent properties which make such materials suitable for A.M. S.M.A. materials such as Ni and Ti are highly suitable in several engineering domains, from bio-medical implants to micro-electromechanical machines. These materials show high sensitivity to the surrounding atmosphere such as light, temperature gradient and humidity. In the clothing industry and the jewellery-based industry, S.M.P.s are used because of their digital light processing behaviour (Zarek et al., 2016). Energy harvesting and actuation applications of these materials are the subjects of ongoing research for 3D nano-fabrication.

The era of Industry 4.0 will have a great inclination towards the application of A.M. with further research and developments on the quality of parts fabricated by A.M. Multi-material A.M. of smart materials is another field of interest that can potentially be explored when establishing factories of the future. Ge et al. (2014) demonstrated the manufacturing of active hinges that have a potential application in self-opening satellite parts, activated by surrounding conditions. In addition, the authors also explored the self-folding/unfolding peculiarity of 3D-printed S.M.P.s as a useful origami. In Industry 4.0, the potential applications include self-assembling fabrications, compact shapes and stimuli-activated mechanisms in a harsh environment. The factories of the future will potentially need smart materials which can be easily programmed and controlled in terms of their properties.

5.3.4 AM FOR RAPID TOOLING

In the purview of developing smart factories, A.M. can be combined with conventional casting methods to produce rapid tooling. A.M. technology can be integrated with casting and be utilised for producing casting patterns, cores and shells (Cheah et al., 2005). S.L.S. and S.T.L. are normally used for the manufacturing of shells and cores using ceramic/sand. In metal casting for making moulds, cores of silica and zirconium are produced by the S.L.S. technique. Das et al. (1998) and Halloran et al. (2011) highlighted the manufacturing of turbine airfoils from investment casting techniques by implementing the Large Area Maskless Photopolymerisation (L.A.M.P.) method. In this, U.V. light is used to cure the suspensions of ceramics powders in its monomer solutions. A.M. makes it convenient to manufacture casting patterns such as polymer pattern by S.T.L., wax pattern by F.D.M., Paper pattern by 3-dimensional printing (3D.P.) and S.L.S. In metal casting of a corresponding metal cast product, shells and core are also shown in Figure 5.11, using 3D.P.

5.3.5 CONCLUSIONS

Industry 4.0, factories of the future, facilitate intelligent factories having high competence and proficient in fabricating quality customised outputs by integrating cyber-physical integration. The recent revolution in information technology has expedited the shift to the inevitable industrial revolution. The nature of the expected industrial revolution primarily depends on the capacities of A.M. High product quality and characteristics can be obtained by executing new materials open for A.M., such as smart materials and metallic components. Novel A.M. technologies provide immense opportunities. However, the design and production challenges are only restrained by the creativity of the individuals.

REFERENCES

Anand, Gautam, and Prateek Saxena. 2016. "A Review on Graphite and Hybrid Nano-Materials as Lubricant Additives." *IOP Conference Series: Materials Science and Engineering* 149 (1). doi:10.1088/1757-899X/149/1/012201.

Balla, Vamsi Krishna, Susmita Bose, and Amit Bandyopadhyay. 2008. "Processing of Bulk Alumina Ceramics Using Laser Engineered Net Shaping." *International Journal of Applied Ceramic Technology* 5 (3): 234–42. doi:10.1111/j.1744-7402.2008.02202.x.

Banack, Trevor M., and Jodi Sherman. 2013. "Trigger Point Injection Indications: Subspecialties: Pain." *Anesthesiology Keywords Review: Second Edition* 20: 579.

Beaman, Joseph, Mukesh Agarwala, Joel Barlow, Harris Marcus, and David Bourell. 2002. "Direct Selective Laser Sintering of Metals." *Rapid Prototyping Journal* 1 (1): 26–36. doi:10.1108/13552549510078113.

Cheah, C. M., C. K. Chua, C. W. Lee, C. Feng, and K. Totong. 2005. "Rapid Prototyping and Tooling Techniques: A Review of Applications for Rapid Investment Casting." *International Journal of Advanced Manufacturing Technology* 25 (3–4): 308–20. doi:10.1007/s00170-003-1840-6.

Concept Laser. n.d. "No Title." http://www.concept-laser.de.

Cormier, Denis, Ola Harrysson, and Harvey West. 2004. "Characterization of H13 Steel Produced via Electron Beam Melting." *Rapid Prototyping Journal* 10 (1): 35–41. doi:10.1108/13552540410512516.

Das, Suman, Martin Wohlert, Joseph J. Beaman, and David L. Bourell. 1998. "Producing Metal Parts with Selective Laser Sintering / Hot Isostatic Pressing PRESSING." *JoM* 50: 17–20.

Dilberoglu, Ugur M., Bahar Gharehpapagh, Ulas Yaman, and Melik Dolen. 2017. "The Role of Additive Manufacturing in the Era of Industry 4.0." *Procedia Manufacturing* 11: 545–54. doi:10.1016/j.promfg.2017.07.148.

Ge, Qi, Conner K. Dunn, H. Jerry Qi, and Martin L. Dunn. 2014. "Active Origami by 4D Printing." *Smart Materials and Structures* 23 (9). doi:10.1088/0964-1726/23/9/094007.

Grigoriev, Alexey, Igor Polozov, Vadim Sufiiarov, and Anatoly Popovich. 2017. "In-Situ Synthesis of Ti2AlNb-Based Intermetallic Alloy by Selective Laser Melting." *Journal of Alloys and Compounds* 704: 434–42. doi:10.1016/j.jallcom.2017.02.086.

Halloran, John W., Vladislava Tomeckova, Susan Gentry, Suman Das, Paul Cilino, Dajun Yuan, Rui Guo, et al. 2011. "Photopolymerization of Powder Suspensions for Shaping Ceramics." *Journal of the European Ceramic Society* 31 (14): 2613–19. doi:10.1016/j. jeurceramsoc.2010.12.003.

Heinl, Peter, Andreas Rottmair, Carolin Körner, and Robert F. Singer. 2007. "Cellular Titanium by Selective Electron Beam Melting." *Advanced Engineering Materials* 9 (5): 360–64. doi:10.1002/adem.200700025.

Herzog, Dirk, Vanessa Seyda, Eric Wycisk, and Claus Emmelmann. 2016. "Additive Manufacturing of Metals." *Acta Materialia* 117: 371–92. doi:10.1016/j. actamat.2016.07.019.

Khoo, Zhong Xun, Joanne Ee Mei Teoh, Yong Liu, Chee Kai Chua, Shoufeng Yang, Jia An, Kah Fai Leong, and Wai Yee Yeong. 2015. "3D Printing of Smart Materials: A Review on Recent Progresses in 4D Printing." *Virtual and Physical Prototyping* 10 (3): 103–22. doi:10.1080/17452759.2015.1097054.

Körner, C. 2016. "Additive Manufacturing of Metallic Components by Selective Electron Beam Melting - A Review." *International Materials Reviews* 61 (5): 361–77. doi:10.10 80/09506608.2016.1176289.

Kruth, J. P., G. Levy, F. Klocke, and T. H. C. Childs. 2007. "Consolidation Phenomena in Laser and Powder-Bed Based Layered Manufacturing." *CIRP Annals - Manufacturing Technology* 56 (2): 730–59. doi:10.1016/j.cirp.2007.10.004.

Kruth, JeanPierre, Peter Mercelis, J. Van Vaerenbergh, Ludo Froyen, and Marleen Rombouts. 2005. "Binding Mechanisms in Selective Laser Sintering and Selective Laser Melting." *Rapid Prototyping Journal* 11 (1): 26–36.

Kumar, S., and J. P. Kruth. 2010. "Composites by Rapid Prototyping Technology." *Materials and Design* 31 (2): 850–56. doi:10.1016/j.matdes.2009.07.045.

Lewis, Gary K, and Eric Schlienger. 2000. "Practical considerations and capabilities for laser assisted direct metal deposition." Los Alamos National Laboratory, Los Alamos, NM and Sandia National Laboratory, Albuquerque, NM.

Lidong, Wang, and Wang Guanghui. 2016. "Big Data in Cyber-Physical Systems, Digital Manufacturing and Industry 4.0." *International Journal of Engineering and Manufacturing* 6 (4): 1–8. doi:10.5815/ijem.2016.04.01.

Liu, Yongkui, and Xun Xu. 2016. "Industry 4.0 and Cloud Manufacturing: A Comparative Analysis." In *ASME 2016 11th International Manufacturing Science and Engineering Conference, MSEC 2016*. Vol. 2. doi:10.1115/MSEC2016-8726.

Malik, Ankit, and Balasubramanian Kandasubramanian. 2018. "Flexible Polymeric Substrates for Electronic Applications." *Polymer Reviews* 58 (4): 630–67. doi:10.1080/ 15583724.2018.1473424.

Optomec. n.d. "No Title." https://optomec.com/.

Otubo, Jorge, and André da Silva Antunes. 2010. "Characterization of 150mm in Diameter NiTi SMA Ingot Produced by Electron Beam Melting." *Materials Science Forum* 643: 55–59. doi:10.4028/www.scientific.net/msf.643.55.

Phenix. 2015. "Phenix Systems." http://www.phenix-systems.com/en/phenix-systems.

Safari, Ahmad. 2001. "Processing of Advanced Electroceramic Components by Fused Deposition Technique." *Ferroelectrics* 263 (1): 45–54. doi:10.1080/00150190108225177.

Sames, W. J, F. A. List, S. Pannala, R. R. Dehoff, and S. S. Babu. 2016. "The Metallurgy and Processing Science of Metal Additive Manufacturing." *International Materials Reviews* 61 (5): 315–60. doi:10.1080/09506608.2015.1116649.

Saxena, Prateek, Michail Papanikolaou, Emanuele Pagone, Konstantinos Salonitis, and Mark R. Jolly. 2020. "Digital Manufacturing for Foundries 4.0." In Tomsett, A. (eds) *Light Metals 2020. The Minerals, Metals & Materials Series* (pp. 1019–1025). Springer, Cham.

Saxena, Prateek, Marie Schinzel, Manuela Andrich, and Niels Modler. 2016. "Development of a Novel Test-Setup for Identifying the Frictional Characteristics of Carbon Fibre

Reinforced Polymer Composites at High Surface Pressure." *IOP Conference Series: Materials Science and Engineering* 149 (1). doi:10.1088/1757-899X/149/1/012124.

Singh, Sunpreet, Seeram Ramakrishna, and Rupinder Singh. 2017. "Material Issues in Additive Manufacturing: A Review." *Journal of Manufacturing Processes* 25: 185–200. doi:10.1016/j.jmapro.2016.11.006.

Strondl, A., M. Palm, J. Gnauk, and G. Frommeyer. 2009. "Microstructure and Mechanical Properties of Nickel Based Superalloy IN718 Produced by Rapid Prototyping with Electron Beam Melting (EBM)." *Materials Science and Technology* 27 (5): 876–83. doi:10.1179/026708309x12468927349451.

Sun, W., D. J. Dcosta, F. Lin, and T. El-Raghy. 2002. "Freeform Fabrication of Ti3SiC2 Powder-Based Structures: Part I - Integrated Fabrication Process." *Journal of Materials Processing Technology* 127 (3): 343–51. doi:10.1016/S0924-0136(02)00284-4.

Tang, Ming, and P. Chris Pistorius. 2017. "Oxides, Porosity and Fatigue Performance of AlSi10Mg Parts Produced by Selective Laser Melting." *International Journal of Fatigue* 94: 192–201. doi:10.1016/j.ijfatigue.2016.06.002.

Thompson, Mary Kathryn, Giovanni Moroni, Tom Vaneker, Georges Fadel, R. Ian Campbell, Ian Gibson , Alain Bernard, et al. 2016. "Design for Additive Manufacturing: Trends, Opportunities, Considerations, and Constraints." *CIRP Annals - Manufacturing Technology* 65 (2): 737–60. doi:10.1016/j.cirp.2016.05.004.

Zarek, Matt, Michael Layani, Ido Cooperstein, Ela Sachyani, Daniel Cohn, and Shlomo Magdassi. 2016. "3D Printing of Shape Memory Polymers for Flexible Electronic Devices." *Advanced Materials* 28 (22): 4449–54. doi:10.1002/adma.201503132.

6 Ensuring Sustainability in Industry 4.0
Implementation Framework

6.1 GUIDELINES FOR ENSURING SUSTAINABILITY IN INDUSTRY 4.0

Sivakumar K., Deepak Mathivathanan, M. Nishal, and Vimal K.E.K.

6.1.1 INTRODUCTION

In recent times, the business processes of industry have been altered due to the technological disturbance created by the rapid change in digitalisation. This is categorised as the fourth industrial revolution or Industry 4.0, where the arrival of contemporary technologies such as Big Data and analytics, blockchain technology, internet of things, augmented reality and rapid prototyping is observed (UNIDO, 2017). These technologies are employed on the assembly line to build a cyber-physical system, where they can create data regarding the activities to be executed on the manufacturing methods in real time and make these data accessible across the organisations (Braccini & Margherita, 2019). In addition to that, these technologies are used to enhance the connectivity among the various elements across the industry, to create a comprehensive and sustainable industrial development (UNDP, 2015; Hidayatno et al., 2019).

The Industry 4.0 concept can be explained in a lot of ways. A simple explanation stated that

> Industry 4.0 deals with the branch of material production through incorporating technologies (Simulation of Things-IoT, Big Data & Analytics) and innovative elements, several devices (Cloud computing, Cyber-Physical Systems (CPS) and Internet of Things (IoT)) and the operational perspectives are addressed/accessed as services and make sure of a persistent communication and relationship
>
> **(Marr, 2018).**

Furthermore, a few experts and researchers defined Industry 4.0 in a different way as 'a new level of value chain organization and product lifecycle management' (Kagermann et al., 2013) or 'collective term for technologies and concepts of chain organization' (Hermann et al., 2015). However, it is different from the definitions given from the organisation's point of view, which state that Industry 4.0 was 'integrating complex machines and physical devices with sensors and networking in the network, the software used to predict, control and plan better results in business and society' (The Industrial Internet Consortium, 2014). In the future, the Industry 4.0 concept might be detected as a competitive strategy, yet with a strong emphasis on value chain optimisation as a consequence of dynamic production and autonomous control (Turkes et al., 2019).

Nowadays, sustainability has turned out to be an essential orientation for organisations because of an increase in population worldwide, pollution, shortages of natural resources and changes in climatic conditions (Boons et al., 2013). Furthermore, sustainability drives the industry to change the traditional factories into smart factories by accepting digital technologies controlling the assembly line, and has become one of the key drivers of Industry 4.0 (Kagermann et al., 2013). However, researchers have discussed the Industry 4.0 applications from the environmental aspects (Braccini & Margherita, 2019). However, how sustainability can be ensured in Industry 4.0 is focused only on by very few researchers. In this chapter, the second part discusses sustainability in Industry 4.0, and then the set of 16 guidelines to ensure sustainability in Industry 4.0 is examined in the third part. The fourth part summarises the outcomes i.e., the benefits and impacts of implementing sustainability in Industry 4.0, and conclusions are presented in the fifth part.

6.1.2 SUSTAINABILITY IN INDUSTRY 4.0

The fourth industrial revolution in manufacturing, which is commonly called Industry 4.0, has shown that manufacturing can work together with emerging technologies to obtain the utmost output with minimal usage of resources, bringing a completely new perspective to the way industries operate. Adolph et al. (2016) stated Industry 4.0 to be a 'German project which merged the manufacturing with information technology'. The consequence of this cooperation leads to the growth of industries that are 'smart' i.e., they are extremely effective in the usage of resources and also they can modify themselves rapidly to fulfil the objectives of management and present-day situations in the industry (Kamble et al., 2018).

Industry 4.0 is required to enhance their overall performance and to govern maintenance with interactions from the surroundings through transforming the standard machines into self-learning and self-aware machines. Also, Industry 4.0 has an objective to construct an open, smart manufacturing platform for the use of an industrial network. The principal requirements of Industry 4.0 are monitoring real-time data, tracing the positions and status of the product, along with maintaining the instructions to master the production processes (Vaidya et al., 2018).

The concept of sustainability was first introduced in 1987 by the United Nations World Commission on Environment and Development. Also, they defined sustainability as 'development that meets the needs of the present without compromising the ability of future generations to meet their own needs' (WCED, 1987). Sustainability is a multi-dimensional concept which incorporates social, economic and environmental dimensions. These three dimensions can overlap, interact and also conflict occasionally. Also, organisations can perform sustainably only when the dimensions of sustainability (i.e., ecological, economic and social) are supported concurrently by the firm (Elkington, 1997; Braccini & Margherita, 2019). Industry 4.0 is anticipated to enhance the worsening values of social, environmental and economic dimensions of sustainability through employing the contemporary technologies of Industry 4.0 and integrating various processes (Stock & Seliger, 2016). The creation of more sustainable industrial values through the participation of Industry 4.0 will be significant in the future. In the past literature studies, this part was mostly credited to the economic and environmental dimensions of sustainability. Apart from the environmental and economic sustainability, Industry 4.0 has enormous potential to understand the creation of sustainable industrial value from the social aspects (Kamble et al., 2018). However, to create such industrial value, some guidelines are to be followed to ensure sustainability in Industry 4.0. In the subsequent section, a brief discussion on the 16 guidelines identified from the existing literature is presented in light of creating better understanding and on how sustainability can be ensured in organisations that are practising Industry 4.0.

6.1.3　Guidelines for Ensuring Sustainability in Industry 4.0

This section focused on the guidelines to make sure that sustainability is in the concept of Industry 4.0. A set of 16 guidelines are identified from the existing literature by reviewing journals from various databases such as SCOPUS, Web of Science and Google Scholars using keywords which include 'guidelines for sustainability in Industry 4.0', 'Industry 4.0 guidelines', 'sustainability in Industry 4.0', 'implementation guidelines for Industry 4.0' individually and with the combination of keywords. With the above search, the related articles are identified, and from further meticulous reviewing, a set of 16 guidelines were listed through the collective effort of the authors. The summary of guidelines is presented in the subsections below.

(i) *Know-how education for the employees*: Industry 4.0 requires content knowledge and how to use the devices connected to it. It also allows an individual to know the activities which can be executed from his or her log-in (Flatt et al., 2016). This requires a specific skill set which has to be learned by training and exercise. The activities cannot be acquired over a single run. Still, a set of prototype testing is needed, the technical difficulties encountered by the employees have to be addressed and training of some sort has to be executed. This sort of testing could help in initiating a smooth transformation from the existing system to Industry 4.0 (Schleipen et al., 2015).

(ii) *Interoperability with standardised module*: The operation is not going to take place within a closed loop, as the current revolution is taking up the system over the internet, which needs interoperability with the other devices (Flatt et al., 2016). To connect with the other systems or devices, either to monitor or execute an order, the module which is going to be designed has to be standardised in such a way that it takes into consideration every other cell module that is used in a different department, and also the work execution has to be taken into account for creating a standardised module for interoperability (Stock & Seliger, 2016).

(iii) *Information transparency*: Information is to be shared with stakeholders and customers to show transparency in the system. Thus a module to reveal the monitoring activity or the completion level of activity with numerous parameters is the current method of showing that the system is stable and sustainable (Hofmann & Rüsch, 2017). The transparency demonstrates the quality of the product, product design, sustainability of the process and environmental impact on the stakeholders. These are being currently demanded by the stakeholders and customers. Also, this helps the system to clearly identify the areas of improvement and also a marketing strategy to cover the customers (Wang et al., 2016).

(iv) *Automation in simpler ways and actionable insights*: Automation is one of the important parts of Industry 4.0, as the automation has to be done first before considering implementing new industry technology (Flatt et al., 2016). Thus, during this process, every organisation has to see further into the future to see what are the requirements and how best they can be achieved effectively, that is, during the implementation of automation for the entire system to be done more simply (Stock & Seliger, 2016), so that any action that needs to be taken in implementing the next level can be converted or upgraded with ease.

(v) *Scalability consideration*: The system has to take into account what size it is going to get converted or take in Industry 4.0 concepts with it. Consider a system that is good at certain areas, when implementing the new system could open up a new option where the system could utilise its resources more effectively (Hofmann and Rüsch, 2017). An entire organisation cannot convert overnight to any new system. Thus scalability has to be considered while doing so that the areas which could have great success in adapting the new technology with ease should be used for a test run of the new system (Wang et al., 2016).

(vi) *Control and visibility in a single digital thread across all operations to ensure transparency*: The control over the system has to be given over a single module thread so that the control of any cell can be done with ease and monitored. The visibility of the system to stakeholders and customers will enhance confidence in the system. The visibility of the process for the suppliers could help in connecting with their system supply schedule for developing upcoming schedules accordingly (Erol et al., 2016).

(vii) *Privacy of machine data*: The privacy of the machine data has to be critical and kept private to certain individuals who are directly in control of the system. This is because machine data for certain processes are confidential and have to be kept for people who have direct access to exchange the machine data for production configuration or design change of the product (Erol et al., 2016). This could be a new design of the product or to develop a new process capability for better quality. When the systems come online, there is a thread that could alter the machine data which could be disruptive to the entire system as to this issue, so the machine data has to be separated and kept for certain members to access only (Mrugalska & Wyrwicka, 2017).

(viii) *Information security management system*: This is an important guideline to be followed in designing an information security management system. As certain information needs to be confidential for the manufacturing system, things like the mixing ratio, how a process is done, the material mix ratio, special treatment processes and other parameters depending on the organisation have to be kept confidential and remain inaccessible to the third party (Zhong et al., 2017). Thus, the device to be connected to the system for control and monitoring from any place has to be encoded and protected from any hacking types of issues taking over the control of the system. It is possible to limit the information provided, and the accessibility also could stop the unauthorised entry or prevent it causing any damage. The new latest technology is to use flash memory with built-in security features that are flexible for smart factories (Gökalp et al., 2017).

(ix) *Safety zone*: To identify the critical areas and remote areas in an organisation to make sure the control and monitoring module developed under the safety zone. These areas have to be identified to know the criticality of the activity performed in those areas and the need for tight security, and the restriction of access for people meaning only required personnel can be given access and monitoring authorisation (Sackey et al., 2017).

(x) *Uniform interface throughout production is elementary*: Interfacing through the production area is vital to implementing Industry 4.0. Still, the interface has to be uniform throughout the production floor. Even the different floors have to be interfaced with uniformly to work with (Flatt et al., 2016). It is highly required in any manufacturing industry, automobile industry or processing industry as they have many shop floors and different processes; these have to be uniformly interfaced for easier control over the process and monitoring (Gökalp et al., 2017).

(xi) *Having simpler integration of plants and machine into a system for easy monitoring*: The integration of machines to the plants using sensors and P.L.C. automation has to be done with simpler integration. Our technology is ever-changing with continuous new improvement; thus, simpler integration will enable the plant and machine to be flexible for any new changes with the technology (Schleipen et al., 2015). Thus, always using simpler integration of plants and machines enables the easy monitoring of the system and any changes or modification or alteration to be performed for improvement can be executed with ease, as the simpler integration enables maintenance activity more effectively and appropriate maintenance can be performed as per requirements.

(xii) *Uniform communication interface within cells*: Industry 4.0 has standardised communication that allows them to carry out scheduling automatically for the system and the supplier, distribution to consumers and governing of orders, which comprise all the production processes and the resources required (Erol et al., 2016). Individual process modules have greater flexibility when compared to the past, and thus it can be combined and even their specific abilities can be used. Cloud could be used for the communication application by Industry 4.0, which could be very effective and flexible to incorporate with the existing module of the system, as this can be used for communicating within and also with the suppliers and consumers (Zhong et al., 2017).

(xiii) *Control device and field device require different communication module which needs to be secure*: Certain devices which have been authorised for controlling the process and machine data within the system requires a communication module which is to be secured in such a way that it cannot be used without authorisation. The field device for certain remote areas and critical areas for monitoring also have to be secured as these data could be vital for an organisation's survival in its field, as it could be its trade secret (Gökalp et al., 2017).

(xiv) *Standardised information system is needed to plug in and work with a new device*: The information system has to be developed in such a manner where important key variables such as control devices, field devices, remote areas, critical areas and other requirements by the industry have to be taken into consideration, while standardising the information system so that the new devices can be engaged with the existing system for work-related activities (Mrugalska & Wyrwicka, 2017). Thus, the system could be accessible from any devices which can be plugged in and the standardised system could enable the protocol on how much accessibility can be allocated to the new device plugged to the system. This covers a lot of concerns as who is trying to gain access to the system (Wang et al., 2016).

(xv) *Feedback system to be retrieved with the help of a standardised system as it has a lot of data to be processed*: Creating the feedback system for the entire process and for all the parameters involved in the production, as well as how the systems are running with or without any glitch in the industry, is an essential part of the standardised system. Thus, any abnormality is to be recorded and accessed at first instance to find the root cause of any unauthorised activity (Funk et al., 2016). The feedback system is crucial as it gives us clearance on how the system is performing and the reliability of the system. A feedback system is quite important for construction of a sustainable system. It also makes the system evolve, with data collected and analysed on how to act in any particular situations in the best suitable way, and to log information about any new undefined activity within the system (Gorecky et al., 2014). This helps the system to make continuous improvements until the sustainable system can be achieved.

(xvi) *Logistics concept designing while adapting Industry 4.0*: Logistics is as important as a complex system where Industry 4.0 shows the significance for cross-organisational logistics concerning end-to-end supply chain transparency, real-time information flows and flexible enhancements, which could lead an organisation to optimise value creations (Hofmann et al., 2017). Ultimately, Industry 4.0 has to be assessed by situations due to the intricate nature of logistics management. Incorporating just-in-time and kanban concepts according to the systems with intelligent bin assessment, and then disposition and production with real-time data could be used while developing the logistics with Industry 4.0 (Franceschini et al., 2003; Hofmann et al., 2017). The above is the level one phase; then comes the second phase delivery and collection based upon the demand run, and last with the goods received via auto-id where the R.F.I.D. tag currently in existence can be incorporated.

These guidelines are provided to help practising managers in accomplishing sustainable development, which leads to improving the performance of Industry 4.0 through attaining the various sustainable benefits and can also be used by industrial managers to identify the sustainability adoption level in Industry 4.0 based on the guidelines followed.

6.1.4   IMPACT OF SUSTAINABILITY IN INDUSTRY 4.0

In recent times, society has become conscious of the impacts on the environment due to industrial value creation, especially after the 'Brundtland Report of the World Commission on Environment and Development' was published (WCED, 1987; Elkington, 1994). The multi-dimensional concept of sustainability should be taken into consideration within the strategic and organisational attention upon the execution of Industry 4.0 because of their interrelationships among the environmental, economic and social aspects (Schulz and Flanigan, 2016). In addition to that, society and policymakers anticipate that organisations in the future will keep on producing the sustainable value to chase the social, ecological and economic goals (Hartmann et al., 2014). The benefits of the Industry 4.0 concept can be unfolded completely when all three dimensions of sustainability generate the value while cautiously balancing the challenges among themselves (Birkel et al., 2019).

A limited number of experimental studies aimed at the sustainability impacts of Industry 4.0 from the perspective of the triple bottom line (T.B.L.) are available in the past literature. Concerning the economic perspective of T.B.L., a lot of studies stated that Industry 4.0 brings about enhancements regarding efficiency and effectiveness, using a novel way to produce goods and optimising the supply chain. Implementation of Industry 4.0 can enhance economic performance through utilising predictive maintenance and predictive analysis, which ultimately assist the organisations in minimising the flaws and mistakes over the assembly line. Likewise, concerning the environmental perspective, Industry 4.0 enabled applications related to energy monitoring, which resulted in a reduction of CO_2 emissions and energy efficiency. The savings in energy within an organisation have a positive impact on its productive capacity. Also, the I.o.T. applications enable the re-use of resources through the remanufacturing process. Further, the data generated by the sensors of the I.o.T. are examined through developing a mathematical model for attaining cost reductions and for handling the remanufactured resources dynamically. Regarding the impact on the social dimension, Industry 4.0 decreased the number of safety incidents and raised the confidence of the employees, which results in the formation of a safer workplace (Braccini & Margherita, 2018).

6.1.5 Conclusion

The study presented in this chapter aimed at providing a better understanding for managers and practising engineers on the sustainability aspects in Industry 4.0. The findings will assist the managers in understanding the guidelines and their impacts on T.B.L. These insights will help managers towards assuring the guidelines for the adoption of sustainability in Industry 4.0. The guidelines considered in this chapter are generic, and thus, they can be used to extend the study with insertion or elimination of a few guidelines. Also, this study explored the impacts of sustainability on Industry 4.0 with a T.B.L. perspective. Furthermore, this chapter is a useful source to build future studies exploring the interlinks between sustainability and Industry 4.0, and it also provides a roadmap for implementing sustainability in Industry 4.0 to any organisation.

REFERENCES

Adolph, Lars. *German Standardization Roadmap: Industry 4.0. Version 2*, Berlin: DIN eV, 2016.

Alhaddi, Hanan. "Triple bottom line and sustainability: A literature review." *Business and Management Studies* 1, no. 2 (2015): 6–10.

Basl, Josef. "Penetration of Industry 4.0 principles into ERP vendors' products and services–a central European study." In *International Conference on Research and Practical Issues of Enterprise Information Systems*, Cham: Springer, (2017): 81–90.

Birkel, H. S., Veile, J. W. et al. "Development of a risk framework for Industry 4.0 in the context of sustainability for established manufacturers." *Sustainability* 11, no. 2 (2019): 384.

Boons, Frank, Carlos Montalvo, et al. "Sustainable innovation, business models and economic performance: An overview." *Journal of Cleaner Production* 45 (2013): 1–8.

Borowy, Iris. *Defining Sustainable Development for Our Common Future: A History of the World Commission on Environment and Development*. Routledge: Brundtland Commission, 2013.

Braccini, A. M., and Margherita, E. G. "Exploring organizational sustainability of industry 4.0 under the triple bottom line: The case of a manufacturing company." *Sustainability* 11, no. 1 (2019): 36.

Braccini, Alessio, and Emanuele Margherita. "Exploring organizational sustainability of industry 4.0 under the triple bottom line: The case of a manufacturing company." *Sustainability* 11, no. 1 (2019): 36.

Elkington, John. "The triple bottom line." *Environmental management: Readings and Cases* 2 (1997).

Elkington, John. "Towards the sustainable corporation: Win-win-win business strategies for sustainable development." *California Management Review* 36, no. 2 (1994): 90–100.

Erol, Selim, Andreas Jäger, et al. "Tangible Industry 4.0: A scenario-based approach to learning for the future of production." *Procedia CIRP* 54 (2016): 13–18.

Flatt, Holger, Sebastian Schriegel, et al. "Analysis of the Cyber-Security of industry 4.0 technologies based on RAMI 4.0 and identification of requirements." In *2016 IEEE 21st International Conference on Emerging Technologies and Factory Automation (ETFA)*, Berlin, Germany, EEE, (2016): 1–4.

Franceschini, F., M. Galetto, A. Pignatelli, and M. Varetto. "Outsourcing: Guidelines for a structured approach." *Benchmarking: An International Journal* 10, no. 3 (2003): 246–260.

Funk, Markus, Thomas Kosch, et al. "Motioneap: An overview of 4 years of combining industrial assembly with augmented reality for industry 4.0." In *Proceedings of the 16th International Conference on Knowledge Technologies and Datadriven Business*, Graz, Austria, (2016): 1–4.

Gökalp, Ebru, Umut Şener, et al. "Development of an assessment model for industry 4.0: Industry 4.0-MM." In *International Conference on Software Process Improvement and Capability Determination*, Cham: Springer, (2017): 128–142.

Gorecky, Dominic, Mathias Schmitt, et al. "Human-machine-interaction in the industry 4.0 era." In 2014 *12th IEEE international conference on industrial informatics (INDIN)*, Porto Alegre, RS, Brazil, IEEE, (2014): 289–294.

Hartmann, Julia, and Sabine Moeller. "Chain liability in multitier supply chains? Responsibility attributions for unsustainable supplier behavior." *Journal of Operations Management* 32, no. 5 (2014): 281–294.

Hermann, Mario, Tobias Pentek, and Boris Otto. "Design principles for Industrie 4.0 scenarios: A literature review. Technische Universitat Dortmund." *Working Paper* (2015): 15.

Hidayatno, Akhmad, Arry Rahmawan Destyanto, et al. "Industry 4.0 technology implementation impact to industrial sustainable energy in Indonesia: A model conceptualization." *Energy Procedia* 156 (2019): 227–233.

Hofmann, Erik, and Marco Rüsch. "Industry 4.0 and the current status as well as future prospects on logistics." *Computers in Industry* 89 (2017): 23–34.

Joshi, Devin K., Barry B. Hughes, and Timothy D. Sisk. "Improving governance for the post-2015 sustainable development goals: Scenario forecasting the next 50 years." *World Development* 70 (2015): 286-302.

Kagermann, Henning, Johannes Helbig, et al. *"Recommendations for Implementing the Strategic Initiative Industrie 4.0: Securing the Future of German Manufacturing Industry"*; final report of the Industrie 4.0 Working Group. Forschungsunion, 2013.

Kamble, Sachin S., Angappa Gunasekaran, and Shradha A. Gawankar. "Sustainable Industry 4.0 framework: A systematic literature review identifying the current trends and future perspectives." *Process Safety and Environmental Protection* 117 (2018): 408–425.

Keeble, Brian R. "The Brundtland report: 'Our common future'." *Medicine and War* 4, no. 1 (1988): 17–25.

Marr, Bernard. "What is industry 4.0." *Here's a Super Easy Explanation for Anyone*. Forbes, September 02, 2018. https://www.forbes. com/sites/bernardmarr/2018/09/02/what-is-industry-4-0-heres-a-super-easy-explanation-for-anyone.

Mrugalska, Beata, and Magdalena K. Wyrwicka. "Towards lean production in industry 4.0." *Procedia Engineering* 182 (2017): 466–473.

Nagasawa, T., C. Pillay, et al. "Accelerating clean energy through industry 4.0: Manufacturing the next revolution." A Report of the United Nations Industrial Development Organization (2017).

Sackey, Samuel Mensah, Andre Bester, et al. "Industry 4.0 learning factory didactic design parameters for industrial engineering education in South Africa." *South African Journal of Industrial Engineering* 28, no. 1 (2017): 114–124.

Schleipen, Miriam, Evgeny Selyansky, et al. "Multi-level user and role concept for a secure plug-and-work based on OPC UA and AutomationML." In 2015 *IEEE 20th Conference on Emerging Technologies & Factory Automation (ETFA)*, Parc Hotel Alvisse, Luxembourg, IEEE, (2015): 1–4.

Schulz, Steven A., and Rod L. Flanigan. "Developing competitive advantage using the triple bottom line: A conceptual framework." *Journal of Business & Industrial Marketing* 31, no. 4 (2016): 449–458.

Stock, Tim, and Günther Seliger. "Opportunities of sustainable manufacturing in industry 4.0." *Procedia CIRP* 40 (2016): 536–541.

Türkeş, Mirela Cătălina, Ionica Oncioiu, et al. "Drivers and barriers in using industry 4.0: A perspective of SMEs in Romania." *Processes* 7, no. 3 (2019): 153.

Vaidya, Saurabh, Prashant Ambad, et al. "Industry 4.0–a glimpse." *Procedia Manufacturing* 20 (2018): 233–238.

Wang, Shiyong, Jiafu Wan, et al. "Towards smart factory for industry 4.0: A self-orga-
       nized multi-agent system with big data based feedback and coordination." *Computer
       Networks* 101 (2016): 158–168.
Zhong, Ray Y., Xun Xu, et al. "Intelligent manufacturing in the context of industry 4.0: A
       review." *Engineering* 3, no. 5 (2017): 616–630.

6.2 CASE STUDIES – SUSTAINING GLOBAL COMPETITIVENESS WITH INDUSTRY 4.0

R. Subhaa, R. Sudhakara Pandian, Leos Safar, and Jakub Sopko

6.2.1 INTRODUCTION

In the digitalised world, small and medium enterprises (S.M.E.s) are facing day-
to-day challenges regarding their sustainable development, especially in the highly
competitive global market environment. Numerous companies are working on digitalisation processes with the aim of implementing the Industry 4.0 concept. Industry
4.0 is the fourth industrial revolution enabled by advances in technology. Hence,
it requires high levels of technical skills and technology implementation strategies
which are still very far away as far as S.M.E.s are concerned. On the other hand,
S.M.E.s play a vital role in any economy and thus it is required to equally and simultaneously develop S.M.E.s along with large enterprises for a balanced development.

The cost of automation, complexity of the technology and non-awareness of new
paradigms of Industry 4.0 have led the S.M.E.s to avoid or postpone implementation
of Industry 4.0 (Cheng-Ju Kuo et al., 2017).

On this concern, this chapter looks at the cases of implementing Industry 4.0 with
the focus on S.M.E.s. This will be useful for strategic planners and designers to help
them identify the challenges and issues of implementing Industry 4.0.

6.2.2 CHALLENGES AND ISSUES OF INDUSTRY 4.0

The Industry 4.0 production model generally requires fast and dynamic decisions
based on online data. This requires complex technology with high computational
efficiency for implementing Industry 4.0. The problem with available technologies
like linear discriminant analysis, Markov analysis and decision trees is that they are
time-consuming, complex and they employ offline data (Cheng-Ju Kuo et al. 2017).
But in the Industry 4.0 scenario, batch production, once done, may not be repeated,
and thus all the data collected will not be needed anymore. And also, even if batch
production continues, the data will again be useless as the more time will be consumed in formulation and analysis and much time has already elapsed. Further, most
of the S.M.E.s are still using traditional machines which need a lot of investment for
the changeover to Industry 4.0. Such S.M.E.s will be reluctant to move to Industry
4.0 unless we provide proven cost-effective solutions to such problems.

The production scenario develops a lot of data like quality data, machine capabilities data, machine health data, maintenance data and production process data. Under
Industry 4.0, such data pose a big storage problem and also in processing and analysing the data. Apart from size, such data need to be retrieved and analysed quickly for
taking any rapid decisions for a faster response for Industry 4.0 (Ateeq & Klaus, 2016).

Often, such data are to be shared with customers or suppliers for the effective functioning of Industry 4.0 and its achievement. These data are to be integrated with technologies for the development of Cyber-Physical Systems (C.P.S.). The systems are to be integrated both widely and deeply for the successful implementation of any technologies. Often, integration is a highly challenging task as it requires major changes in data structure and also requires methods for transformation of data.

The mass customisation concept of Industry 4.0 needs the processes to be flexible to cope with the changing demand and shorter product life-cycle (Ateeq & Klaus, 2016). This flexibility should also be quick and high-quality, with more effective and efficient processes. This further requires flexibility in all departments of the company. Above all, these flexibility and adaptations are to be done in a cost-effective manner.

6.2.3 TECHNOLOGIES OF INDUSTRY 4.0

This section discusses various technologies needed for the implementation of Industry 4.0. As Industry 4.0 is technology-centric and requires more sophisticated technologies, this section elaborates some of the technologies like internet of things (I.o.T.), cyber-physical systems (C.P.S.), cloud computing and Big Data analytics specific to Industry 4.0.

6.2.3.1 Internet of Things (I.o.T.)

I.o.T.-enabled manufacturing is the initiative for smart manufacturing and it uses sensors, networks, optimisation algorithms and data integration techniques for collecting and analysing data (Zhong et al., 2017). On-demand use and efficient sharing of technologies enable the Industry 4.0 system to be a smart one. I.o.T. offers advanced connectivity among physical objects, services and systems and is able to share and exchange data among these. I.o.T. also uses web-based technologies to integrate heterogenous systems and gives access to a selected subset of data. It has link technologies, protocols and other services required for connecting digital devices (Chen et al., 2014). Further, I.o.T. has real-time data collection and uses R.F.I.D. (Radio Frequency IDentification) and wireless technologies for tracing and monitoring the components of manufacturing systems. In this, R.F.I.D. tags and readers are attached to movable objects that may be humans or products and use R.F.I.D. for identification and interconnectivity, so that they become intelligent and hence can be called smart products. Such tags when tied to inbound logistics can be traced and monitored and real-time data collection can also be done. Such data can be used to detect any discrepancies and also automatically correct or alert or initiate the required rectifications. This makes the products smart products. This helps to work out shop floor activities in a quicker and more flexible response manner and also allow data collection for real-time decision-making.

6.2.3.2 Cyber-Physical Systems (C.P.S.)

Cyber-physical systems (C.P.S.) are the mechanism by which physical objects are made to exchange information through software and other linked embedded systems (Bagheri et al., 2015). It basically involves interaction of different components of a system through exchange of information. For instance, the intelligent manufacturing systems in Sweden and the U.S.A. have C.P.S.-associating holons, agents and function blocks. This has the advantages of implementing decentralised or cloud environment, maximised flexibility and advanced condition monitoring.

6.2.3.3 Cloud Manufacturing

Cloud manufacturing is the internet-based computing service and it does service distribution and sharing of data. It has intelligent capability management and manufacturing cloud service management. It uses technologies like cloud computing, I.o.T., virtualisation methods and service-oriented technology. The ideal cloud computing has the characteristics of on-demand service, broad network access and resource pooling (Zhong et al., 2017). The deployment modes may be public, private, community and hybrid. The primary challenges of deploying cloud computing are security issues and load balancing. This is applied in data visualisation in a cloud manufacturing shop floor and smart cloud manufacturing using the I.o.T.

6.2.3.4 Big Data Analytics

Big Data represents an enormous dataset that is mostly unstructured and is required for real-time analysis (Chen et al., 2014). The intrinsic challenges of Industry 4.0 are to deal with Big Data, derive useful information and react to it (Oliff & Liu, 2017). The Big Data characteristics are (i) voluminous data, (ii) data of many varieties and of different data structures, (iii) variability, (iv) availability at different speeds i.e., velocity of data differs and (v) veracity (Hilbert, 2016). In this regard, Big Data storage, acquisition, management and analysis are big tasks. The data quality mainly depends on accuracy, consistency, completeness and redundancy of the data (Chen et al., 2014). Thus, Big Data technology requires ways to detect data quality and repair the damaged dataset. Big Data require data encryption for the security and safety of the data. And also, the data are to be categorised or clustered to identify the required and crucial data for continuous real-time monitoring of the processes, and the relevant data are to be taken to another physical system for essential decision-making.

6.2.4 Case Studies Based on Industry 4.0 Technologies

This section discusses the case studies carried out in various industries based on Industry 4.0 technologies. It also addresses their implementation processes, problems and issues encountered in implementation.

6.2.4.1 Cases on Neural Network Technologies

The machines of a spring factory, in this case (Cheng-Ju Kuo et al., 2017), are fixed with sensors in key points of the machines and data are collected, parsed and categorised to get useful data that are important and also suitable for analysis, then a neural network model is designed for high prediction accuracy and high operating speed. The major advantage of this model is the low cost that makes it suitable for S.M.E.s. This model enables the beginning of the automatic machine prediction leading to Industry 4.0.

6.2.4.2 Case Studies on I.o.T. Technologies

This case study given by Zanella and Vangelista (2014) addresses the usage of I.o.T. technologies for smart administration of cities making it a 'Smart City'. The case study describes Padova Smart City implemented in the city of Padova, Italy and collaborates with the city municipality. In this system, the street lights are fitted with sensors and devices for measuring air temperature, humidity, carbon footprint, vibrations and noise. It also checks the correct operation of light intensity depending upon weather conditions, etc. These data are linked to the Padova municipality servers so that useful information regarding traffic congestion and pollution control is generated, monitored and analysed and the necessary actions are triggered so that the city will be a smart city using Industry 4.0 technologies. This system uses the I.o.T. technologies like web-based devices, back-end servers, database management system, enterprise resource planning systems (E.R.P.), gateways and link layer technologies. Figure 6.1 shows the system architecture of Padova Smart City.

6.2.4.3 Cases on Big Data Technologies

In any organization, data are a critical and important asset, and often the problem with data is its volume is unmanageable and thus it poses a Big Data problem (Stojanovic et al., 2016). The model-driven diagnostics are complex and time-consuming and ultimately require Industry 4.0 for ubiquity sensing and proactive responses using Big Data technologies. This case study, given by Stojanovic et al. (2016), deals with the usage and implementation of Big Data technologies.

The clustering-based anomaly detection is applied in this case for fault detection of the component (i.e., a propeller fan). The deviations in the components are inherent due to the production process. The abnormal deviations itself are to be identified and immediate actions are to be taken to rectify and avoid quality problems. This real-time monitoring requires Big Data, as it requires a large memory size. For detecting this anomaly, the deviations are clustered to identify the normal and abnormal deviations. This requires Big Data technologies and this case study uses Hadoop slaves. It uses YARN for cluster and AMBARI for Hadoop management. For investigating, the performance changes with the size of the data; a different

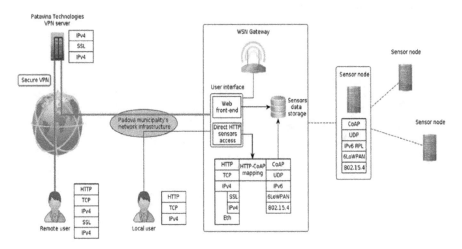

FIGURE 6.1 System architecture of Padova smart city (Courtesy: Zanella and Vangelista, 2014).

set-up is created by doing clustering many times. This approach is thus robust and changes the way the quality control is done. This paves the way for highly efficient quality management in the context of Industry 4.0.

Another case study by Oliff and Liu (2017) uses data mining technologies for an efficient decision-making system. This case is conducted in an S.M.E. that manufactures washing machines and tumble dryers of different models. Here, data mining is used to identify patterns of failure and the knowledge of the pattern is useful in making informed automatic decisions. The dataset is in archived form and consist of textual descriptions of faults and also actions taken. Based on these datasets, a model is designed to produce a set of rules for the decision-making system. Data mining software, WEKA (Waikato Environment for Knowledge Acquisition), that is available as freeware is used for input data and this makes the system a low-cost and easily implementable technology.

6.2.4.4 Cases on Industrial Wireless Network (I.W.N.) Technologies

Industrial Wireless Networks (I.W.N.) has the advantages of low cost, flexibility and easy implementation and plays an important part in Industry 4.0 and implementation for a smart factory (Li et al., 2017). This case study of Li et al. (2017) explains the steps of implementation of I.W.N. to achieve smart manufacturing using cloud and Big Data. The main goal of Industry 4.0 is mass customisation and this case details the principle and operation of smart factories to achieve this goal.

Figure 6.2 shows the main components of the I.W.N. models. It has a smart factory with R.F.I.D. tagged raw products, smart sensors, smart machines and smart robots, a conveyor system, I.W.N. system management, smart terminals and a private industrial cloud.

The principle of operation of the I.W.N. model is as follows:

i. User provides input of their preferences and choices to the webpage, which is then loaded to the server

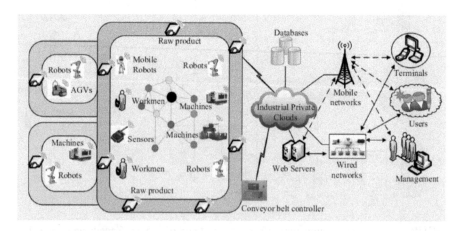

FIGURE 6.2 Architecture of case study based on I.W.N. (Courtesy: Li et al., 2017).

ii. The private cloud builds the machine processing program, related conveyance requirements and issues instructions to smart machines and workers

iii. The other related information is fed to the management system for necessary validation and decision-making.

Here, I.W.N. provides a link for the raw products, machines, conveyance and networks to communicate and enable the realisation of Industry 4.0 and smart factory.

6.2.4.5 Industrial Internet of Things (I.I.o.T.) Technologies

The case studies that involve technologies based on Industrial Internet of Things (I.I.o.T.) and are commercially available are presented here.

This case study ('Smart cylinder and Industry 4.0', 2018) presents a smart gas cylinder that uses I.I.o.T. to self-monitor the gas levels and does inventory management without need for human intervention. In this case, a mobile application mentioning the stack is developed based on Bluetooth LE wireless technology. This application will scan the gas cylinder while entering the warehouse and from the measured gas level, it evaluates the time to stock out. It reorders the gas cylinder, considering the lead time. The timely reorder enables effective and efficient inventorying, reducing the cost of inventory and reducing the risk of stock outs. Figure 6.3 shows the working architecture of a smart gas cylinder. This application will be useful for an industry that uses gas cylinders as raw materials or other energy supply and enables order replenishment and hence implementation of Industry 4.0 using a wireless network. These smart cylinders can also be incorporated into a smart supply chain where replenishment and delivery are done without human intervention and will avoid stock outs and also minimise inventory.

The case study about Air Separation Units (A.S.U.) ('Air separation units and Industry 4.0', 2018), in which nitrogen and oxygen are separated for industrial uses, show a lot of variation in demand. A.S.U. also require very tight quality control. Such units require a lot of time for shifting of orders and may incur heavy losses

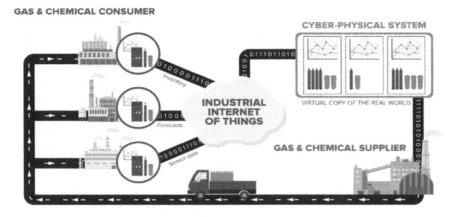

FIGURE 6.3   Smart gas cylinder supply chain case study (Courtesy: 'Smart Cylinder and Industry 4.0', 2018).

when quality issues arise. Industry 4.0 provides an opportunity for A.S.U. by using the Industrial Internet of Things (I.I.o.T.) and expert systems. This model enables remote operation of independent A.S.U. operators at a low cost. This provides access to real-time performance of the plant and also makes adjustments to the quality requirements of the customers. These remotely located A.S.U. operating centres not only monitor the performance, they also adjust the operation.

Many organisations take initiatives to reduce energy consumption, reduce CO_2 emissions and install technologies that could bring about these advantages. But the predicted savings are often not realised due to poor monitoring of the energy consumption. This case study aims to avoid energy waste with an Industry 4.0 approach ('Avoid energy waste', 2018). In Industry 4.0, such cases are regularly monitored using I.I.o.T. and any discrepancies are immediately taken care of. The steps of the approach are initially: (i) new installations are checked for the actual savings against predicted saving/advantages through completion management system (C.M.S.) and then, (ii) monitor energy consumption and analyse deviations from predicted or expected performance using a Collaborative Energy Efficiency Programme (C.E.E.P.)

Often, in industries, energy costs are higher and also least interfered with due to unavailability of technologies to monitor energy usage. Hence, this case study ('Energy management', 2018) deals with energy management using Industry 4.0 techniques and technologies to avoid such energy loss. The case study has used C.P.S. to capture real-time data on the door operation of a direct fired radiant tube heater for heating requirements of the shop floor space. The measured temperature, set points and gas consumption enable C.P.S. to make decisions with the goal of reducing or minimising energy waste. This, when integrated with the controller allows the system to stop operation when the doors are opened, and prevent high gas consumption during cold days.

6.2.4.6 Logistics Optimisation Technologies

In this case study, a dairy company categorised as an S.M.E., located in the Madurai region, India is considered, because of its vision and openness towards new trends.

A dairy is a place where cows' milk is processed for daily consumption. A dairy is located on an exclusive farm or as a combined farm along with the harvesting of milk. The building or farm area where milk is harvested is often called a milking parlour. The farm area where milk is stored in bulk tanks is known as the farm's milk house. Milk is then transported by a lorry to a dairy plant where the raw milk is further processed and prepared for commercial sale. There are different processes involved in the dairy company, not only to produce pasteurised milk, but also a few allied products like ghee, butter, cheese, ice-cream and yogurt. Since the company is increasing production on year-to-year basis, it is of the utmost importance to implement new concepts in order to achieve higher competitiveness, reduce costs and improve quality.

In the actual entry process at the dairy plant to test and accept the raw milk, the vehicles which carry the milk wait in a queue for their turn. This consumes a lot of time every day due to the large quantities of raw milk from different village farms and from different regions that are being taken to the single point. It is also not recommended to keep the raw milk in the actual outside temperature for a long time. The milk should be kept at 4°C immediately after milking. Otherwise there is a chance for an increase in the bacterial count that will affect the quality of the final products. To avoid such problems, in this case, we use Industry 4.0 techniques suitable for minimising the duration of the wait for and transport of the raw milk. Also, suitable solutions for each stage after milking for both incoming material and products at the plant are recommended, with the aim of making monitoring from the plant easier.

In this case study, we initially calculated the maximum perimeter around the plant, and the corresponding village farms where milking is done were noted down. The number of locations where the incoming raw milk and outgoing final products will be stored was decided based on the facilities available in total from eight to twelve locations and based on the expansion planned by the dairy plant.

A model was proposed to allocate the dairy farms in different regions based on various constraints. A simple heuristic technique was developed to analyse the optimal distance for transporting both incoming and outgoing materials from the dairy plant. The calculation of waiting time results obtained should serve as a recommendation for the selected dairy plant in this case study.

6.2.5 SUMMARY AND FINAL REMARKS

Industry 4.0 converts a traditional factory to a smart factory by integrating individual components of the production systems so that they can communicate with each other to make informed decisions, and produce self-aware, self-monitoring production components. The main characteristics of a smart factory are that it is small-lot and has mass customisation i.e., customised customer demands (Wang et al., 2015). The smart factory needs to be flexible and react quickly to changes. It should be well-integrated with other elements of the supply chain like suppliers, distributors, customers, etc. Above all, it should be cost-effective and should be of high quality. And also, many varieties of models are produced simultaneously in the conveyor-type production system.

Under Industry 4.0 and its allied technologies like I.o.T., Big Data analytics, I.W.N., I.I.o.T., cloud computing, etc. a traditional industry is converted to a smart factory. Though the smart factory has yet to be realised widely, we have the necessary technologies and strategies for the implementation of these technologies. Thus, a smart factory as a result of Industry 4.0 can be flexible at meeting the needs and demands of each and every customer in better quality and also at a very much cheaper price. Such days are not far away, so we have to prepare and get ready for Industry 4.0.

REFERENCES

Air separation units and industry 4.0, downloaded from www.trcontrolsolutions.com/case-studies Accessed on 20 Dec. 2018.

Andrea, Zanella, and Lorenzo Vangelista. 2014. "Internet of things for smart cities." *IEEE Internet of Things Journal* 1(1): 22–32.

Avoid energy waste with an industry 4.0 approach, downloaded from www.trcontrolsolutions.com/case-studies Accessed on 20 Dec. 2018.

Bagheri, Behrad, Shanhu Yang, Hung-An Kao, and Jay Lee. 2015. "Cyber-physical systems architecture for self-aware machines in industry 4.0 environment." *IFAC-PapersOnLine* 48(3): 1622–1627.

Chen. M., S. Mao, and Y. Liu. 2014. "Big data: A survey," *Mobile Networks and Applications* 19(2): 171–209.

Cheng-Ju Kuo, Kuo-Cheng Ting, Yi-Chung Chen, Don-Lin Yang, and Hsi-Min Chen. 2017. "Automatic machine status prediction in the era of industry 4.0: Case study of machines in a spring factory." *Journal of Systems Architecture* 81:44–53.

Energy management and industry 4.0, downloaded from www.trcontrolsolutions.com/case -studies Accessed on 20 Dec. 2018.

Hilbert, M. 2016. "Big data for development: A review of promises and challenges." *Development Policy Review* 34: 135–174.

Khan Ateeq, and Turowski Klaus. 2016. "A survey of current challenges in manufacturing industry and preparation for industry 4.0." In *Proceedings of the First International Scientific Conference "Intelligent Information Technologies for Industry (IITI'16), Advances in Intelligent Systems and Computing."* edited by Abraham et al., doi:10.1007/978-3-319-33609-1_2.

Li, X., D. Li, J. Wan et al. 2017. "A review of industrial wireless networks in the context of industry 4.0." *Wireless Networks* 23: 23. doi:10.1007/s11276-015-1133-7.

Ljiljana, Stojanovic, Marko Dinic, Nenad Stojanovic, and Aleksandar Stojadinovic. 2016. "Big-data- driven anomaly detection in industry (4.0): An approach and a case study." *IEEE International Conference on Big Data (Big Data)*, 1647–1652. doi: 10.1109/BigData.2016.7840777.

Oliff, Harley, and Ying Liu. 2017. "Towards industry 4.0 utilizing data-mining techniques: A case study on quality improvement." *Procedia CIRP* 63: 167–172.

Smart cylinder and industry 4.0, downloaded from www.trcontrolsolutions.com/case-studies Accessed on 20 Dec. 2018.

Wang, S., JiafuWan, Di Li, and Chunhua Zhang. 2015. "Implementing smart factory of industrie 4.0: An outlook." *International Journal of Distributed Sensor Networks.* doi:10.1155/2016/3159805.

Zhong, R. Y., Xun Xu, Eberhard Klotz, and Stephen T. Newman. 2017. "Intelligent manufacturing in the context of industry 4.0: A review." *Engineering* 3: 616–630.

6.3 MODELLING THE INTERRELATIONSHIP OF FACTORS ENABLING

AGILE-INDUSTRY 4.0: A DEMATEL APPROACH

Srijit Krishnan, Sumit Gupta, and K. Mathiyazhagan

6.3.1 INTRODUCTION

An increasing number of initiatives aimed at improving operations have ushered in a change in the role of manufacturing (Yusuf & Adeleye, 2002). Globalisation demands the use of various practices to sustain a business in a dynamic market. Traditional manufacturing firms are product-oriented and focused more on quantity than quality. Today, firms adopt the 'customer is king' (5 Reasons Why the Customer is King, 2018) principle to sustain business with the current dynamic market trends. Agile manufacturing works on this principle to make machines adaptable to several needs and deliver different types of end products. Agile aims to make production systems as modular as possible, enabled by smart or intelligent components of Industry 4.0 cooperating with one another. This enables establishing various set-up combinatory configurations so it can be reusable, smart and is able to be shared among the various production processes in real time. In order to support this agile flexibility, it is essential to make use of cyber-physical systems and other Industry 4.0 components like internet of things, big data and cloud computing.

This chapter reviews the literature available in order to comprehend how agile manufacturing can be coupled with the future trend of Industry 4.0. Common factors that can have an effect on either one of them have been identified and the DEMATEL approach is used to prioritise these factors based on data obtained from experts in industry and academia. The results obtained are discussed and a managerial implication is provided.

6.3.2   LITERATURE REVIEW

Agile manufacturing is described as a 'system of doing business' (Groover, 2019). Agile manufacturing has been defined by Groover (2019) as one that has the ability to thrive in a competitive environment where continuous and sometimes unforeseen changes are observed by being able to introduce new products into a rapidly changing market through an enterprise-level manufacturing strategy. Agile manufacturers exhibit certain characteristics that differentiate them from traditional manufacturers. Agile manufacturers are capable of *adjusting to changes*. As a result of a dynamic market, any *physical and human resources are rapidly reconfigured to adapt* accordingly. Value is given to knowledge and any innovation to enforce an agile organisation is encouraged. The management is responsible for providing resources that personnel need. There is devolution of authority and there is 'a climate of mutual responsibility for joint success' (Groover, 2019). Another characteristic is 'cooperation to enhance competitiveness'. The object of organisations is to bring products into the market as rapidly as possible, and to sustain this, partnering with other firms, even with competing companies, may be required. A major agile characteristic is to 'enrich the customer'. The modern manufacturing firm concentrates on satisfying customers' desires. For this to happen, the products' pricing depends on what value the solution proposed has to the customer, rather than the cost of manufacturing.

Industry 4.0 is a concept that doesn't have a concrete definition. According to the research by Zhong et al. (2017),

> Industry 4.0 combines embedded production system technologies with intelligent production processes to pave the way for a new technological age that will fundamentally transform industry value chains, production value chains, and business models. In the context of Industry 4.0, manufacturing systems are updated to an intelligent level.

Industry 4.0 was defined by Vaidya et al. (2018) as 'a new level of organization and control over the entire value chain of the life cycle of products'. This industry era incorporates Big Data, internet of things and cyber-physical systems to enhance the manufacturing environment and make it smart. Smart manufacturing, as we call it nowadays, relies on information communication technology. Converting existing regular machines into self-learning systems is necessary to improve overall management of performance and maintenance coupled with interaction with the surroundings (Vaidya et al., 2018).

In today's fast-paced dynamic market, an increased need for individualised products, implying a shift from mass production to mass customization, is observed. It is essential to cater to the continuous consumer demands that vary at a rapid pace. It is necessary to exercise agile concepts along with Industry 4.0-era technology to create a more integrated environment. One of the major drivers of agile manufacturing is technology. In this chapter I propose the use of Industry 4.0 technology and concepts to enable agility in manufacturing industries. This is achieved by 'the Agile factory implemented as Cyber-physical Human System (CPHS)' (Gunasekaran et al., 2018). Scheuermann et al. (2015) created such a C.P.H.S. using smart factory hardware along with an accepted manufacturing framework, D.E.S.C. The article by Scheuermann et al. (2015) suggests that the Industry 4.0-era agile factory is user-centric whereby customers are enabled to influence assembly processes during runtime and receive live status updates. This means that such a factory is 'user centric as the workers and customers serve as sensors and actuators'. In this chapter, the factors that affect an Industry 4.0 agile factory have been identified and their description provided in Table 6.1.

6.3.3 DECISION-MAKING TRIAL AND EVALUATION LABORATORY

6.3.3.1 DEMATEL Approach

This method has proven to be helpful in solving complex problems. The DEMATEL approach was developed by Geneva Battelle Memorial Institute in 1976. This approach is advantageous compared to others because it evaluates the power of relationships between variables identified in the study (Chen, 2016). Demonstrated below are the stages adopted to conduct the DEMATEL method (Bacudio et al., 2016). This method, used frequently in the literature, is useful in assessing the performance of complex systems. It is used in cases such as creating a model for sustainable consumption and production (Luthra et al., 2017); evaluation of supply chain performance in hospitals (Supeekit et al., 2016); defining the success factors critical in emergency management (Zhou et al., 2017); evaluation of CRM partners (Büyüközkan et al., 2017).

In this research, the methodology adopted by Yüksel et al. (2017) is followed. The following steps are carried out to obtain the results.

Step 1: Each respondent k generates an initial direct relation matrix. Equation (6.1) illustrates the respondents' matrix. Based on expert opinion, the matrix is created. The respondents analyse the factors and their relationship between them and makes the pairwise comparisons as per the scale provided in Table 6.2.

$$A_k = \begin{bmatrix} 0 & \cdots & a_{1nk} \\ \vdots & \ddots & \vdots \\ a_{n1k} & \cdots & 0 \end{bmatrix} \qquad (6.1)$$

TABLE 6.1

Factors to Enable Agile Industry 4.0

Criteria	Sub-Criteria	Definition	Supporting Literature
Decision-Making	Top management commitment, C1	Management is required to understand the importance of implementing new practices to sustain in the market.	(Singh et al., 2007)
	Decentralisation, C2	Delegated authority throughout the organisation and to all levels of management.	(Hermann et al., 2016)
	Production methodology, C3	Method by which a product is built/made, decided by an engineering team.	(Dangayach & Deshmukh, 2001; Quintana, 2005)
Economic	Set-up Cost, C4	A requirement to get the manufacturing plant equipped and ready for production,	(Kloukinas & Rotondi, 2012)
Technology	Use of I.T. and its integration, C5	Essential requirement in successfully implementing connectivity between employees and between machinery.	(Faller & Feldmúller, 2015)
	Manufacturing flexibility, C6	Ability to make amendments to manufacturing processes based on current trends in the market.	(Zhong et al., 2017)
	Real-time capabilities, C12	Ability to instantaneously respond and convey information.	(Hermann et al., 2016)
Social	Employee training and involvement, C7	A multi-skilled labour force is a necessity in today's technological advanced environment	(Singh et al., 2007)
	Social dimension of sustainability, C13	Ability to develop and implement a community accepted self-sustaining methodology without jeopardising the potential of future generations to meet their needs.	(Gupta et al., 2016; Mani et al., 2016)
	Environmental conservation, C9	Greener practices must be followed for conservation of the environment.	(Yusuf & Adeleye, 2002)
	Competitive pressure, C10	A level of market competition that affects business performance.	(Yusuf & Adeleye, 2002)
Quality	Preventive maintenance, C8	A way of preventing any mishaps from happening to manufacturing systems by adopting preventive measures.	(Kamble et al., 2018)
	Continuous improvement, C11	An incremental improvement of efforts or practices conducted in the industry.	(Gunasekaran et al., 2018)

Step 2: This step identifies the relationship among the elements. An initial influence matrix is calculated (Table 6.3).

Step 3: Equation (6.2) is used to normalise the direct relation matrix. 'b_{ij}' in this equation returns a value between 0 and 1.

TABLE 6.2

Likert Scale for Survey

Linguistic Description	Value
No Effect	0
Very Low Effect	1
Low Effect	2
High Effect	3
Very High Effect	4

$$B = \left[b_{ij} \right]_{nxn} = \frac{A}{\max \sum_{j=1}^{n} a_{ij}} \qquad (6.2)$$

Step 4: Using Equation (6.3), the total relation matrix is developed. In this equation, 'C' represents this matrix and 'I' is the identity matrix.

$$C = \left[c_{ij} \right]_{nxn} = B(I - B)^{-1} \qquad (6.3)$$

Step 5: Equation (6.4) and (6.5) are taken into consideration to calculate the prominence (D + E) and cause-effect (D − E) values.

$$D = \left[d_{ij} \right]_{nx1} = \left[\sum_{j=1}^{n} c_{ijij} \right]_{nx1} \qquad (6.4)$$

$$E = \left[e_{ij} \right]_{1xn} = \left[\sum_{j=1}^{n} c_{ijij} \right]_{1xn} \qquad (6.5)$$

Step 6: Creation of prominence and cause-effect graph.

6.3.3.2 Application of DEMATEL

See Tables 6.3 to 6.6 and Figures 6.4 and 6.5.

6.3.4 RESULTS AND DISCUSSION

From the factors identified, the M.C.D.M. technique provides the values of prominence (D + E) and cause-effect (D − E) which are shown in Table 6.6. The prominence graph shown in Figure 6.4 represents the correlation extent among the factors. That is, it illustrates how far each factor affects another. Production methodology – C3 has the highest prominence of 9.74 followed by manufacturing flexibility – C6 with 9.62 in value. The value 9.74 observed in case of C3 indicates that its influence on other variables is very high, followed by C6. On the contrary, it is noticed that

TABLE 6.3
Average Initial Influence Matrix

	C1	C2	C3	C4	C5	C6	C7	C8	C9	C10	C11	C12	C13
C1	0	3.6	2.4	3	2.8	4	0.8	1.8	3	4	3	4	3
C2	3.8	0	3.2	4	1.6	3	2	3	3.4	3.8	3.2	2	0.6
C3	3.8	3.2	0	3.8	4	4	3	3	3	3.8	3	4	1.8
C4	4	1.4	4	0	4	4	0.8	1	0	4	2	4	1.6
C5	3	3.2	4	4	0	4	3	4	0	3	3.2	4	4
C6	3.4	3	4	4	4	0	2.4	3	1.2	3	2	4	0.8
C7	3.4	3.2	3.2	1.8	3.2	4	0	3.6	0.8	4	4	0.8	3.8
C8	1.8	1	3.2	2	4	3.8	4	0	1.6	3	3.8	1.8	0.6
C9	3.2	1	0.6	3	0	3	0	3	0	2.8	3	0.8	4
C10	4	1.8	3.2	1	3	3.8	3.2	1.8	3.2	0	3.8	1	3.2
C11	2.8	3.2	4	1	4	4	3	3	3.8	4	0	1.8	2.8
C12	1	3.2	3.8	3.8	4	4	2	3.2	0.8	3	1.8	0	1
C13	3.4	0.8	1.8	1.6	1	0	3.2	1.8	3.2	3	3	1.2	0

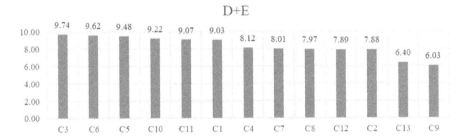

FIGURE 6.4   Prominence graph.

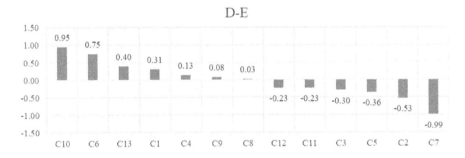

FIGURE 6.5   Net cause/effect graph.

environmental conservation – C9 and social dimension of sustainability – C13 have values of 6.03 and 6.40 respectively, indicating very low influence on other factors, as well as on the system itself.

Figure 6.5 represents the net cause-effect graph (D – E) for the factors. If the D – E value is greater than zero, the factor is considered a cause factor. This implies that those factors with D – E values of less than zero are effect factors. Competitive pressure – C10, Manufacturing flexibility – C6, Social dimension of sustainability – C13,

TABLE 6.4

Normalised Matrix B

	C1	C2	C3	C4	C5	C6	C7	C8	C9	C10	C11	C12	C13
C1	0.0000	0.0865	0.0577	0.0721	0.0673	0.0962	0.0192	0.0433	0.0721	0.0962	0.0721	0.0962	0.0721
C2	0.0913	0.0000	0.0769	0.0962	0.0385	0.0721	0.0481	0.0721	0.0817	0.0913	0.0769	0.0481	0.0144
C3	0.0913	0.0769	0.0000	0.0913	0.0962	0.0962	0.0721	0.0721	0.0721	0.0913	0.0721	0.0962	0.0433
C4	0.0962	0.0337	0.0962	0.0000	0.0962	0.0962	0.0192	0.0240	0.0000	0.0962	0.0481	0.0962	0.0385
C5	0.0721	0.0769	0.0962	0.0962	0.0000	0.0962	0.0721	0.0962	0.0000	0.0721	0.0769	0.0962	0.0962
C6	0.0817	0.0721	0.0962	0.0962	0.0962	0.0000	0.0577	0.0721	0.0288	0.0721	0.0481	0.0962	0.0192
C7	0.0817	0.0769	0.0769	0.0433	0.0769	0.0962	0.0000	0.0865	0.0192	0.0962	0.0962	0.0962	0.0913
C8	0.0433	0.0240	0.0769	0.0481	0.0962	0.0913	0.0962	0.0000	0.0385	0.0721	0.0913	0.0433	0.0144
C9	0.0769	0.0240	0.0144	0.0721	0.0000	0.0721	0.0000	0.0721	0.0000	0.0673	0.0721	0.0192	0.0962
C10	0.0962	0.0433	0.0769	0.0240	0.0721	0.0913	0.0769	0.0433	0.0769	0.0000	0.0913	0.0240	0.0769
C11	0.0673	0.0769	0.0962	0.0240	0.0962	0.0962	0.0721	0.0721	0.0913	0.0962	0.0000	0.0433	0.0673
C12	0.0240	0.0769	0.0913	0.0913	0.0962	0.0962	0.0481	0.0769	0.0192	0.0721	0.0433	0.0000	0.0240
C13	0.0817	0.0192	0.0433	0.0385	0.0240	0.0000	0.0769	0.0433	0.0769	0.0721	0.0721	0.0288	0.0000

TABLE 6.5
Total Relation Matrix C

	C1	C2	C3	C4	C5	C6	C7	C8	C9	C10	C11	C12	C13
C1	0.3077	0.3223	0.3662	0.3409	0.3627	0.4289	0.2506	0.3041	0.2698	0.4224	0.3560	0.3424	0.2889
C2	0.3819	0.2338	0.3715	0.3504	0.3290	0.4001	0.2662	0.3192	0.2710	0.4080	0.3515	0.2914	0.2312
C3	0.4373	0.3521	0.3598	0.3988	0.4360	0.4837	0.3323	0.3703	0.2960	0.4693	0.4007	0.3818	0.2984
C4	0.3697	0.2604	0.3779	0.2531	0.3697	0.4042	0.2335	0.2655	0.1850	0.3950	0.3097	0.3287	0.2404
C5	0.4134	0.3465	0.4434	0.3955	0.3447	0.4736	0.3324	0.3846	0.2287	0.4459	0.3985	0.3770	0.3357
C6	0.3878	0.3173	0.4092	0.3692	0.3998	0.3525	0.2903	0.3356	0.2291	0.4089	0.3408	0.3520	0.2451
C7	0.3950	0.3226	0.3938	0.3196	0.3828	0.4394	0.2438	0.3516	0.2321	0.4346	0.3907	0.2812	0.3152
C8	0.3222	0.2491	0.3586	0.2931	0.3673	0.3993	0.3028	0.2428	0.2176	0.3727	0.3500	0.2736	0.2219
C9	0.2816	0.1850	0.2255	0.2478	0.2060	0.2958	0.1596	0.2417	0.1413	0.2907	0.2646	0.1906	0.2403
C10	0.3815	0.2732	0.3641	0.2811	0.3499	0.4064	0.2912	0.2924	0.2663	0.3180	0.3612	0.2636	0.2879
C11	0.3924	0.3304	0.4182	0.3152	0.4064	0.4518	0.3158	0.3501	0.3014	0.4446	0.3120	0.3099	0.3039
C12	0.3122	0.2991	0.3793	0.3414	0.3743	0.4099	0.2644	0.3179	0.2025	0.3793	0.3117	0.2410	0.2284
C13	0.2881	0.1856	0.2516	0.2187	0.2279	0.2381	0.2294	0.2214	0.2145	0.2985	0.2699	0.1971	0.1599

TABLE 6.6

Prominence and Cause/Effect Values

	D	E	D + E	D – E
C1	4.67092	4.362781	9.033701	0.308139
C2	3.677391	4.20508	7.882471	−0.527689
C3	4.719163	5.016548	9.73571	−0.297385
C4	4.124766	3.992845	8.117611	0.131921
C5	4.556443	4.91986	9.476303	−0.363418
C6	5.18368	4.437653	9.621333	0.746026
C7	3.512256	4.502405	8.014661	−0.99015
C8	3.997287	3.971092	7.96838	0.026195
C9	3.055261	2.970427	6.025688	0.084834
C10	5.087813	4.136738	9.224552	0.951075
C11	4.417221	4.652052	9.069273	−0.234831
C12	3.830201	4.061416	7.891617	−0.231216
C13	3.397325	3.000827	6.398153	0.396498

Top management commitment – C1, Setup cost – C4, Environmental conservation – C9 and Preventive maintenance – C8 are cause challenges arranged from the highest to the lowest, with the highest value 0.95 attained by C10 and the lowest positive being C8, with a value of 0.03. On the other hand, Employee training and involvement – C7, Decentralisation – C2, Use of I.T. and its integration – C5, Production methodology – C3, Continuous improvement – C11 and Real-time capabilities – C12 are effect factors arranged from the most negative value of −0.99 attained by C7 to the least negative of −0.23 attained by C12. This implies that competitive pressure – C10 is the primary cause factor and Employee training and involvement – C7 being the most affected factor.

6.3.5 CONCLUSION

The research in this chapter presents factors that enable manufacturing organisations to adopt Industry 4.0 agile concepts. Thirteen factors that enable this combined adoption in manufacturing organisations have been identified through literature and brainstorming session with experts. The DEMATEL method was used to analyse the factors by identifying their prominence and cause-effect values.

The production methodology is identified as the most prominent and influencing factor for the adoption of Industry 4.0-era agile manufacturing practices. It is essential to recognise the method of production adopted in manufacturing industries. We know the main production methods are job, batch and mass production. In the fast-paced market of today, where the customer is the decider of how long a product will survive in the market, a smart methodology needs to be implemented. We know that agility comes from customisation of products and services according to customers' specifications, and with flexible Industry 4.0-era technology, it is possible to provide for these demands. In order to survive in a competing market, customisation is key.

In a traditional manufacturing firm, customisation is a characteristic of job production methodology, but to cater to the masses, customisation of products on a mass scale needs to be considered. This will forge strong relationships with customers leading to their return in the future. Mass customisation also reduces costs for storage and scrap quantity because nothing is produced until an order is received. This means inventories of unsold goods and raw materials are eradicated.

A limitation of such mass customisation methodology is achieving mass production efficiency. A highly flexible production technology needs to be implemented, and that can be time-consuming and expensive.

This research can go deeper by using hybrid tools such as grey-DEMATEL or fuzzy-DEMATEL to consider any uncertainties that arise from collecting the responses. Identifying more factors that impact implementing Industry 4.0 agile practices would further enhance the research. Furthermore, a framework model to implement this Industry 4.0-era agility in manufacturing firms may be developed.

REFERENCES

5 Reasons Why the Customer is King. (2018, August 13). Retrieved from Volom: https://volom.com/5-reasons-why-the-customer-is-king/.

Bacudio, L. R., Benjamin, M. F. D., Eusebio, R. C. P., Holaysan, S. A. K., Promentilla, M. A. B., Yu, K. D. S., & Aviso, K. B. (2016). Analyzing barriers to implementing industrial symbiosis networks using DEMATEL. *Sustainable Production and Consumption*, 7(8), 57–65. doi:10.1016/j.spc.2016.03.001.

Brauers, W. K. M., & Zavadskas, E. K. (2006). The MOORA method and its application to privatization in a transition economy. *Control and Cybernetics*, 35(2), 445–469.

Büyüközkan, G., Güleryüz, S., & Karpak, B. (2017). A new combined IF-DEMATEL and IF-ANP approach for CRM partner evaluation. *International Journal of Production Economics*, 191(October 2016), 194–206. doi:10.1016/j.ijpe.2017.05.012.

Chen, I. S. (2016). A combined MCDM model based on DEMATEL and ANP for the selection of airline service quality improvement criteria: A study based on the Taiwanese airline industry. *Journal of Air Transport Management*. doi:10.1016/j.jairtraman.2016.07.004.

Dangayach, G. S., & Deshmukh, S. G. (2001). Manufacturing strategy literature review and some issues. *International Journal of Operations and Production Management*, 21(7), 884–932. doi:10.1108/01443570110393414.

Faller, C., & Feldmúller, D. (2015). Industry 4.0 learning factory for regional SMEs. *Procedia CIRP*, 32(Clf), 88–91. doi:10.1016/j.procir.2015.02.117.

Groover, M. P. (2019, March 8). *Agile Manufacturing.* Retrieved from BrainKart: https://www.brainkart.com/article/Agile-Manufacturing_6450/.

Gunasekaran, A., Yusuf, Y. Y., Adeleye, E. O., Papadopoulos, T., Kovvuri, D., & Geyi, D. G. (2018). Agile manufacturing: An evolutionary review of practices. *International Journal of Production Research*, (November). doi:10.1080/00207543.2018.1530478.

Gupta, K., Laubscher, R. F., Davim, J. P., & Jain, N. K. (2016). Recent developments in sustainable manufacturing of gears: A review. *Journal of Cleaner Production*, 112, 3320–3330. doi:10.1016/j.jclepro.2015.09.133.

Hermann, M., Pentek, T., & Otto, B. (2016). Design principles for industrie 4.0 scenarios. *Proceedings of the Annual Hawaii International Conference on System Sciences, 2016–March*, 3928–3937. doi:10.1109/HICSS.2016.488.

Kamble, S. S., Gunasekaran, A., & Gawankar, S. A. (2018). Sustainable Industry 4.0 framework: A systematic literature review identifying the current trends and future perspectives. *Process Safety and Environmental Protection*, 117, 408–425. doi:10.1016/j.psep.2018.05.009.

Kloukinas, C., & Rotondi, D. (2012). General challenges and an IoT @ work perspective.

Luthra, S., Govindan, K., & Mangla, S. K. (2017). Structural model for sustainable consumption and production adoption—A grey-DEMATEL based approach. *Resources, Conservation and Recycling, 125*, 198–207. doi:10.1016/j.resconrec.2017.02.018.

Mani, V., Gunasekaran, A., Papadopoulos, T., Hazen, B., & Dubey, R. (2016). Supply chain social sustainability for developing nations: Evidence from India. *Resources, Conservation and Recycling, 111*, 42–52. doi:10.1016/j.resconrec.2016.04.003.

Quintana, R. (2005). A production methodology for agile manufacturing in a high. *International Journal of Operations & Production Management, 18*(5), 452–470. doi. org/10.1108/01443579810206127.

Scheuermann, C., Verclas, S., & Bruegge, B. (2015). Agile factory-an example of an Industry 4.0 manufacturing process. *Proceedings - 3rd IEEE International Conference on Cyber-Physical Systems, Networks, and Applications, CPSNA 2015, 2008*, 43–47. doi:10.1109/CPSNA.2015.17.

Singh, R. K., Garg, S. K., & Deshmukh, S. G. (2007). Interpretive structural modelling of factors for improving competitiveness of SMEs. *International Journal of Productivity and Quality Management, 2*(4), 423. doi:10.1504/IJPQM.2007.013336.

Supeekit, T., Somboonwiwat, T., & Kritchanchai, D. (2016). DEMATEL-modified ANP to evaluate internal hospital supply chain performance. *Computers and Industrial Engineering, 102*, 318–330. doi:10.1016/j.cie.2016.07.019.

Vaidya, S., Ambad, P., & Bhosle, S. (2018). Industry 4.0 - a glimpse. *Procedia Manufacturing, 20*, 233–238. doi:10.1016/j.promfg.2018.02.034.

Yüksel, S., Dinçer, H., & Emir, Ş. (2017). Comparing the performance of Turkish deposit banks by using DEMATEL, Grey Relational Analysis (GRA) and MOORA approaches. *World Journal of Applied Economics, 3*(2), 26–47. doi:10.22440/wjae.3.2.2.

Yusuf, Y. Y., & Adeleye, E. O. (2002). A comparative study of lean and agile manufacturing with a related survey of current practices in the UK. *International Journal of Production Research, 40*(17), 4545–4562. doi:10.1080/00207540210157141.

Zhong, R. Y., Xu, X., Klotz, E., & Newman, S. T. (2017). Intelligent Manufacturing in the Context of Industry 4.0: A Review. *Engineering, 3*(5), 616–630. doi:10.1016/j.eng.2017.05.015.

Zhou, X., Shi, Y., Deng, X., & Deng, Y. (2017). D-DEMATEL: A new method to identify critical success factors in emergency management. *Safety Science, 91*, 93–104. doi:10.1016/j.ssci.2016.06.014.

6.4   DEVELOPMENT OF A NOVEL FRAMEWORK FOR A DISTRIBUTED MANUFACTURING SYSTEM PROCESS FOR INDUSTRY 4.0

Ramakurthi Veera Babu, Vijaya Kumar Manupati,
Nikhil Wakode, and M.L.R. Varela

6.4.1   INTRODUCTION

Current manufacturing enterprises focus on distributed manufacturing environments as a result of challenges brought by globalisation, the customer's customised requirements for their products, outsourcing and simultaneous movement of skilled labour. In order to cater for the characteristics of today's manufacturing functions i.e., make to order with a reduced product life-cycle, advanced technology and customization, the functionalities of a distributed network of manufacturing have the potential to achieve strong interconnection to facilitate the above-mentioned requirements with an efficient collaborative environment.

Achieving efficient collaborative distributed manufacturing to capture the competitive advantage in geographically located enterprises has become a major challenge. To ensure a strong competitive market, industries have been transforming their approaches with a level of designing and manufacturing to fit quantity flexibility without compromising the level of customisation.

6.4.2 LITERATURE REVIEW

The discussion in this section has been focused on the detailed literature of distributed manufacturing systems confined to various modules i.e., ontology, interoperability, multi-agent systems, simulation and telefacturing systems which are mentioned in the following sections.

In conventional manufacturing, the raw material is collected from the supplier's side and then processed in centralised locations to deliver the finished products to the end-user, whereas in the Distributed Manufacturing (D.M.) environment, due to its various flexible resources, the products are transferred to the nearest customer's location to process and deliver. This method is achievable, cost-effective and highly efficient, and can be adapted to different manufacturing paradigms. Therefore, it has been identified as one of the latest growing technologies from the upcoming manufacturing perspective by the United Nations Industrial Development Organization (UNIDO, 2013) and the World Economic Forum's Meta-Council on Emerging Technologies (Klaus, 2011).

Rauch et al. (2017) discussed the plethora of literature that is already available to emphasise the several advantages of the D.M. environment over a centralised manufacturing system. Some of them had greater flexibility to satisfy local customer demand, improved sustainability, reduction in logistics costs and lesser delivery times. However, Rauch et al. (2018) developed a structure of network models for the distributed manufacturing network for smart and agile mini-factories. They also explored the need for a distributed manufacturing structure for achieving sustainability. Woern et al. (2017) presented a D.M. platform for flexible products and discussed their technical feasibility and economic viability in the manufacture of commercial fibres for household purposes. Kendrick et al. (2017) described the strategies to realise decentralised manufacture through hybrid manufacturing platforms.

Adoption of D.M. network modelling in anticipation of a competitive environment made manufacturing systems more complex due to its distinct feature of experts (process engineers, material engineers, system engineers, production managers, etc.) working from various domains such as interoperability, information sharing, product life-cycles in system modelling, etc. The common ontology for gaining knowledge (ideas, things, facts, etc.) and forming relations among them is not sufficient due to the variety of systems and their different configurations. Hence, Zaletelj et al. (2018) proposed a basic ontology formulation theory along with directive principles to provide an environment where the model can be sustained over time. Ontology is considered one of the best methods to support D.M. modelling in relating to physical realisation, operations management and modelling of the system.

Ontology is a technique of presenting an object of expertise that determines relationships and classification of concepts. According to Chungoora and Young (2011), ontology is a representation of a shared conceptualisation of a specific interest by a group of users, whereas foundation ontology signifies the fundamentals on which models can be built. If that ontology is a meta-model, a model is built over another model described as the higher layer. However, if the same foundational ontology is a meta over a meta-model which has a self-description, ontologies can be built, but not guarantee consistency. Consistency needs to be achieved from conceptual dynamics where it captures the temporal dynamic changes of a model and its environment and life-cycle management, according to Zaletelj et al. (2018). According to Kadiri et al. (2015), interoperability is the systematic usage and application to legislation and laws of an ontological reference model to mitigate risk arising due to perturbations from diverse processes and capabilities. Sureephong et al. (2008) defined ontologies as a crucial contribution to knowledge management by providing a superior way of representing knowledge and supporting the development of reclaimable and distributable knowledge bases. Huang et al. (2019) proposed an ontology framework which consists of generation of data and evolution of data for allocating the manufacturing data in mould production processes in the intelligent manufacturing environment.

Young et al. (2009), in his project I.M.K.S., demonstrated interoperable manufacturing knowledge sharing in inter-companies across the product life-cycle with the aid of reference ontologies. To achieve the resemblance in the conflicting distributed manufacturing system, considering interoperability is important. Agostinho and Goncalves (2015) stated that for mutually exchanging information, the most challenging part is coordination and integration among the enterprises. Panetto et al. (2013) stated that for sharing information during coordination, ontology is the mode for interoperability of all the applications. Interoperability is defined as the ability of the system to exchange and understand the shared information among itself. Manupati et al. (2016) described telefacturing as an approach where enterprises exchange information with a centralised network to control various tasks in a system of enterprises, machines, shop floors and suppliers collectively. In their work, the interaction between different business's information has been discussed as one of the critical tasks. Diep et al. (2007) stated that the European Interoperability Framework (E.I.F.) defined interoperability into three layers: the technical layer deals with the link between the connected network and their objectives, the semantic layer deals with validating the ability to understand the objective of the exchanged data that the application is designed to perform and the organisational layers deal with administration goals and processes to combine organisations that are willing to exchange information. Open standards for extendable mark-up language (X.M.L.) and Ethernet will help to overcome the issues of technical interoperability difficulties that can be eliminated with the use of connected networks and X.M.L.

Manupati et al. (2016) developed a mobile-agent based approach in a networked manufacturing environment for the integration of process planning and scheduling. With a contract net protocol-based negotiation method the efficiency and effectiveness of the system have been improved. Mishra et al. (2016) proposed a cloud-based multi-agent architecture for effective planning and scheduling of distributed manu-facturing. With the introduction of eleven autonomous agents, the framework has been designed and it can deal with the internal and external interferences by taking into consideration the short- and long-term prospects of the manufacturing system. Li et al. (2018) developed a multi-agent-based architecture for a distributed manu-facturing system where the shared contract net protocol approach has been used to schedule the jobs independently. Sustainability parameters are of great importance and need to be tested for their effects on the current manufacturing environment. Dmitry et al. (2019) presented a multi-agent deep learning approach for the simulta-neous optimisation of time and energy in a distributed routing system. In their work, a distributed routing approach integrated with machine learning in both communi-cation networks and physical systems is tested to simultaneously optimise the travel time of the routed entities and energy consumption.

6.4.3   A PROPOSED NOVEL FRAMEWORK FOR TELEFACTURING DISTRIBUTED PROCESS FOR INDUSTRY 4.0

The telefacturing concept emerged from the extension of the distributed manufac-turing environment concept where the hubs (common workplace) are allocated to the connected enterprises for processing jobs collaboratively in a single worksta-tion. With the motive of improving the 3Cs i.e., communication, coordination and collaboration of enterprises located in geographically distributed companies, the telefacturing system is designed in the context of Industry 4.0. Moreover, with this system, new kinds of approaches e.g., transportation coordination, worker skill level and flexibility in the working environment can be greatly improved.

In this section, a novel distributed process framework for Industry 4.0 has been proposed and is shown in Figure 6.6. Here, the framework can be conceptualised into three functional blocks. First, the user service level, in which a real-time model has been employed to accommodate information, data processing from the customer to allow the analysis of the collected data and information, in particular, to manage the flow of operations in hubs (environment where workers are engaged in work) cen-trally. Second, the control service level, where the customer's orders will be centrally managed to decide distribution/assigning of tasks for producing product parts in each hub globally. In the final layer i.e., the application service level, the actual product is being produced in the hubs located near to the skilled labour accommodation. In each functional unit information, (bi-directional) flow occurs via the internet of things, Big Data analytics and cloud computing to optimise the processing and logis-tic parameters. In the next section, the detailed explanation of each layer is discussed.

6.4.4   DISCUSSION ON PROCESSING OF TELEFACTURING-BASED DISTRIBUTED SYSTEM

The four pillars of distributed manufacturing are circulated universally i.e., enter-prises, machines, manufacturing lines, plants where they need to work collabora-tively to achieve a strong linkage between them. Achieving efficient collaborative

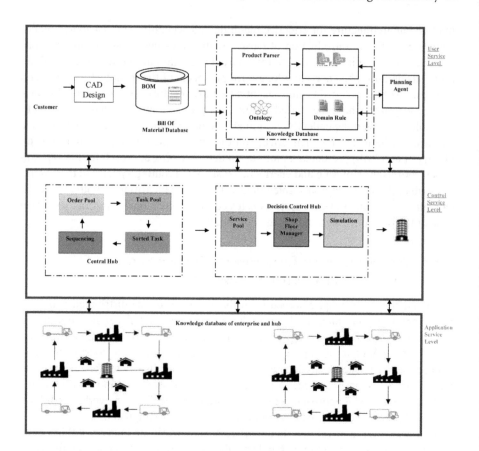

FIGURE. 6.6   Proposed novel telefacturing system framework for Industry 4.0.

distribution manufacturing to capture the competitive advantage in the globally distributed enterprises has become the major challenge. To enrich the competitiveness
in the market, industries are looking for mass production without losing customisation of the product. Therefore, attaining a competitive advantage over others in the
era of Industry 4.0 requires an effective and efficient approach. One such method/
approach is detailed in the section that follows.

6.4.4.1   User Service Level

The first level of the proposed framework is the user service level, presenting the concept of an integrated environment that represents the highly distributed manufacturing
requirement and resources. Figure 6.6 shows the total conceptual model of the user service level make-up framework. Finding a feasible solution for a service scheme specified by the enterprise user and the customer user is organised into four main stages.

1. The customer provides a manufacturing request based on his requirement
   using a C.A.D. environment file
2. The product parser is used to verify and parse the manufacturing request

3. After formulating the queries of series, one can access the integrated knowledge database for any query from the list of feasible manufacturing resources

4. A planning agent suggests the desired service plan based on the time, money and other limitations by considering the given manufacturing request and available resources.

There are two types of customers present in the manufacturing environment i.e., customer users and enterprise users. Customer users are those users who do not have their infrastructure to provide services for the respective order task. The enterprise users are those which themselves have an infrastructure and are capable of further processing or providing other services. In the user service level, the system takes service requests from a service customer and depends on the requirements of the particular customer which include cost, quality constraints and the product specifications that are defined in the form of many personalised projects. The enterprise user or customer user service request depends on a C.A.D. environment such as SolidWorks, Autodesk Inventor, etc. given by the user to formulate the overall specification of the service request and the respective specification is retrieved from the bill of material file (B.O.M.) based on the product requirement. Thereafter, the generated B.O.M. and C.A.D. files are sent to the product parser. Using a web portal, the product parser carries the direct product parsing, then the file is converted into an X.M.L. file with the proposed request given by the user. These requests are represented in the ontology using a predefined data object which converts the variety of received data into the form of one common language i.e., an X.M.L. file. After a service request given by the user level is verified by the product parser, the verified X.M.L. file is pushed into the central knowledge database. After that, the verified file is part of the integrated domain knowledge.

In the telefacturing domain, to arrange various tasks in a systematic manner, globally circulated enterprises have to exchange their information with hubs situated in nearby stations. Here, the information is related to the variety of data related to the product and business processes among the enterprises that hubs have mutually exchanged. This step is the most crucial and difficult for the D.M. platform, as to transfer the data it is necessary to formulate one common language to solve the interoperability and communication issues. Therefore, we introduced ontology for converting information into knowledge and then transferred it into different file formats, out of which it has the compatibility to convert the knowledge into the widely used extensible markup language (X.M.L.) that can transfer the diversified information without any interruption and can mutually exchange mass amounts of data with each other.

In this module, the planning agent examines the information received from individual elements; here we named these Task ID and Resource ID according to the functionalities and their characteristics. In order to examine each task and resource ID, the planning agent queries the resource's owner, availability information, etc. with already available X.M.L. files with the World Wide Web. After a mutual agreement between the planning agent and the task/resource agent, a service plan is constructed and then finalised in the central hub.

6.4.4.2   Control Service Level

The second level in the proposed framework in Figure 6.6 is defined as a control service level. The control service level is divided into two major sections, viz. the Central Hub and the Decision Control Hub.

6.4.4.2.1   Central Hub

Customers have different demands related to their specific products and after placing the order for the different products, the information related to the respective product is transferred into the order pool. After the orders are placed in the order pool, they are decomposed into separate tasks in the task pool.

Orders from various customers are placed in the order pool to complete the order. All the orders are decomposed into individual tasks. For completing one order various tasks are incorporated based on the services given by the enterprises. From the very first stage of design, manufacturing, testing, packaging, processing, etc., all the tasks are included in the respective product order. Various enterprises are available to perform those tasks, and the task pool looks for the available services offered by enterprises. Considering all the factors – the efficiency, reliability of the network, service time, service cost logistics cost, processing cost, etc. – the enterprise's choice is selected. The tasks are scheduled and sequenced amongst the enterprises according to the priorities and requirements of the product. The procedure is repeated for different orders depending upon the customer's requirements.

6.4.4.2.2   Decision Control Hub

After scheduling and sequencing of the task, the decision control hub takes the decision on which enterprises to choose for the respective task, depending upon certain factors. The enterprises are usually small- and medium-scale enterprises and have limited infrastructure for certain types of services. Each enterprise can provide a group of services to the customers from the service pool. The service pool is used for providing different services to the customer with the help of enterprises. These services may be related to design, R&D, manufacturing, processing, packaging, etc. Based on the individual services, the enterprises perform the tasks assigned to them within the specified time. Enterprises provide these services to their customers or clients. The shop floor manager plays an important role in performing a variety of administrative duties in planning, development and implementation to maintain pool facilities. Before giving the services to the enterprises the shop floor manager tries to simulate the particular process whether it is correct or not. If the respective order task is showing the correct result, then the process is carried out into the next step. The order task services are given to the enterprises based on the order specification like design, R&D, manufacturing, processing, packaging, etc. If something went wrong once, the cycle will continue from the order pool until the required result is not shown.

6.4.4.3   Application Service Level

In the application service level, allotment and coordination is the superior part among all geographically distributed environments for mutually exchanging information and performing multiple tasks from the hubs. To achieve the desired outcome, proper identification of the hub and allocation of a task is necessary. In the allocated hub facility, information and knowledge of the product and shared geographic enterprises right from the design of a product to manufacturing is well defined. However, in general, the enterprises are situated in rural areas, while factory plants could set up a hub in a location closer to the residential areas. The cost of the land in that area is cheaper compared to the cost of the land in the urban area. A hub is defined as a facility where workers are engaged in a wide variety of areas of production, depending on their skills. Besides, the workers train their co-workers by making them skilled labourers rather than semi-skilled. Depending on market demand, hubs are willing to share their skills considering the competitiveness which are coming shortly. Every time market demand fluctuates, resources try to respond to these rapid changes to satisfy customer demand. In the future, small hubs will convert into big operating hubs where the infrastructure associated with these hubs can also turn to be a common workplace for several factories, and they can also collaborate with the other enterprises where they can share their information for meeting the actual customer demand. This collaborative service support system is capable of carrying out customer requests through the manufacturing information network (application service layer) through virtual agents. It provides enterprises with resources such as processes that are highly expensive and too complicated for them.

The application service level also a part of the manufacturing information network where it can carry out customer requests through the collaborative service support system. Several advantages can be achieved with this kind of platform. First, it can perform highly expensive and complicated processes provided by enterprises with resources. Second, it is useful for making a simple manufacturing process using a distributed process by integrating enterprises which are situated in various geographical areas. Consequently, the skilled workers who are disabled may have a chance to work in hubs by training their co-workers in the hub, which is one of the cost issues for traditional/centralised manufacturing units. Subsequently, these trained workers may be used as skilled workers in nearby enterprises as and when they require.

In a knowledge database of enterprises and hubs, the information related to product resources, supplier-related information, bills of material (B.O.M.), engineering drawings, process plans and enterprise information is maintained. The tasks related to the selection of potential enterprises, quality control services, profits, valuation of manufacturing capabilities of the enterprises, product status and all the primary knowledge of enterprises and task evaluations are kept in the hub database. Logistics play a crucial role in a distributed manufacturing environment, as the enterprises are distributed geographically. Logistics are used every time when there is a need to transfer from one enterprise to another when assembling a product, distributing subparts in the assembly and transferring finished parts and inventories. If any issue arises in the logistics of the whole network, the processing of the order may be stopped. A robust logistics network between enterprises enhances the flexibility to adapt to any aberrations or unpredicted events related to orders as well as enterprises. The logistics costs and logistic time are critical while solving a problem related to networked manufacturing environments to get optimal solutions.

6.4.5 Conclusions

Distributed manufacturing gains its advantage in the current environment due to its flexibility to carry out different types of tasks effectively and efficiently. Still, there is a need to fully utilise the benefits of Industry 4.0 and its advantages. Hence, current distributed manufacturing needs to adjust and widen its current functionalities for greater benefits. In this chapter, we try to establish a novel framework that can be further generalised not only to the manufacturing systems context, but also to many different applications. Here, the discussion on different functionalities and the processes used in the proposed framework are detailed. In future work, one can use the proposed framework in any of the real-life problems in distributed platforms for an effective process.

REFERENCES

Agostinho, Carlos, and Ricardo Jardim-Goncalves. "Sustaining interoperability of networked liquid-sensing enterprises: A complex systems perspective." *Annual Reviews in Control* 39 (2015): 128–143.

Arsenjev, Dmitry, Dmitry Baskakov, and Vyacheslav Shkodyrev. "Distributed ledger technology and cyber-physical systems. Multi-agent systems. Concepts and trends." In *International Conference on Computational Science and Its Applications*, pp. 618–630. Springer, Cham, 2019.

Chungoora, Nitishal, and Robert Ian Marr Young. "The configuration of design and manufacture knowledge models from a heavyweight ontological foundation." *International Journal of Production Research* 49, no. 15 (2011): 4701–4725.

Diep, Daniel, Christos Alexakos, and Thomas Wagner. "An Onology-based interoperability framework for distributed manufacturing control." In *2007 IEEE Conference on Emerging Technologies and Factory Automation (EFTA 2007)*, pp. 855–862. IEEE, 2007.

El Kadiri, Soumaya, Walter Terkaj, Esmond Neil Urwin, Claire Palmer, Dimitris Kiritsis, and Robert Young. "Ontology in engineering applications." In *International Workshop Formal Ontologies Meet Industries*, pp. 126–137. Springer, Cham, 2015.

Huang, Chengxi, Hongming Cai, Lida Xu, Boyi Xu, Yizhi Gu, and Lihong Jiang. "Data-driven ontology generation and evolution towards intelligent service in manufacturing systems." *Future Generation Computer Systems* 101 (2019): 197–207.

Kendrick, Blake A., Vimal Dhokia, and Stephen T. Newman. "Strategies to realize decentralized manufacture through hybrid manufacturing platforms." *Robotics and Computer-Integrated Manufacturing* 43 (2017): 68–78.

Klaus, Schwab. "The global competitiveness report 2011–2012." *World Economic Forum*. Geneva, Switzerland, 2011.

Li, Kai, Tao Zhou, Bo-hai Liu, and Hui Li. "A multi-agent system for sharing distributed manufacturing resources." *Expert Systems with Applications* 99 (2018): 32–43.

Manupati, Vijaya Kumar, Goran D. Putnik, Manoj Kumar Tiwari, Paulo Ávila, and Maria Manuela Cruz-Cunha. "Integration of process planning and scheduling using mobile-agent based approach in a networked manufacturing environment." *Computers & Industrial Engineering* 94 (2016): 63–73.

Mishra, Nishikant, Akshit Singh, Sushma Kumari, Kannan Govindan, and Syed Imran Ali. "Cloud-based multi-agent architecture for effective planning and scheduling of distributed manufacturing." *International Journal of Production Research* 54, no. 23 (2016): 7115–7128.

Panetto, Hervé, and Joe Cecil. "Information systems for enterprise integration, interoperability and networking: Theory and applications." (2013): 1–6.

Rauch, Erwin, Patrick Dallasega, and Dominik T. Matt. "Distributed manufacturing network models of smart and agile mini-factories." *International Journal of Agile Systems and Management* 10, no. 3–4 (2017): 185–205.

Rauch, Erwin, Marco Unterhofer, and Patrick Dallasega. "Industry sector analysis for the application of additive manufacturing in smart and distributed manufacturing systems." *Manufacturing Letters* 15 (2018): 126–131.

Sureephong, Pradorn, Nopasit Chakpitak, Yacine Ouzrout, and Abdelaziz Bouras. "An ontology-based knowledge management system for industry clusters." In *Global Design to Gain a Competitive Edge*, pp. 333–342. Springer, London, 2008.

UNIDO. "Industrial development report 2013. Sustaining employment growth: The role of manufacturing and structural change." (2013).

Woern, Aubrey, and Joshua Pearce. "Distributed manufacturing of flexible products: Technical feasibility and economic viability." *Technologies* 5, no. 4 (2017): 71.

Young, R. I. M., A. George Gunendran, Nitishal Chungoora, Jennifer A. Harding, and Keith Case. "Enabling interoperable manufacturing knowledge sharing in PLM." In *Proceedings of the Sixth International Conference on Product Life Cycle Management PLM09*, 2009.

Zaletelj, Viktor, Elvis Hozdić, and Peter Butala. "A foundational ontology for the modelling of manufacturing systems." *Advanced Engineering Informatics* 38 (2018): 129–141.

Index